D1386808

AS/A-LEVEL YEAR 1
STUDENT GUIDE

EDEXCEL

Geography

WITHDRAWN

Tectonic processes and hazards
Landscape systems, processes
and change

Cameron Dunn

HODDER
EDUCATION
AN HACHETTE UK COMPANY

Hodder Education, an Hachette UK company, Blenheim Court, George Street, Banbury, Oxfordshire OX16 5BH

Orders

Bookpoint Ltd, 130 Park Drive, Milton Park, Abingdon, Oxfordshire OX14 4SB

tel: 01235 827827

fax: 01235 400401

e-mail: education@bookpoint.co.uk

Lines are open 9.00 a.m.–5.00 p.m., Monday to Saturday, with a 24-hour message answering service. You can also order through the Hodder Education website: www.hoddereducation.co.uk

© Cameron Dunn 2016

ISBN 978-1-4718-6315-8

First printed 2016

Impression number 5 4 3 2 1

Year 2020 2019 2018 2017 2016

Cover photo: Kevin Eaves/Fotolia

Typeset by Integra Software Services Pvt Ltd, Pondicherry, India

Printed in Slovenia

Hachette UK's policy is to use papers that are natural, renewable and recyclable products and made from wood grown in sustainable forests. The logging and manufacturing processes are expected to conform to the environmental regulations of the country of origin.

Contents

■Getting the most from this book

Exam tips

Advice on key points in the text to help you learn and recall content, avoid pitfalls, and polish your exam technique in order to boost your grade.

Knowledge check

Rapid-fire questions throughout the Content Guidance section to check your understanding.

Knowledge check answers

1 Turn to the back of the book for the Knowledge check answers.

Summaries

■ Each core topic is rounded off by a bullet-list summary for quick-check reference of what you need to know.

ⓔ 6/6 marks awarded. The answer to Part (a) is correct for 1 mark. Abrasion would be an alternative correct answer. The student correctly identifies the maximum and minimum years from Figure 1 for Part (b)(i) and shows these in their answer, as well as working out the difference between them correctly scoring 2 marks. Mark schemes for questions such as this have an 'acceptable' range of correct answers, usually ±5% around the precise answer. Notice that in Part (b)(ii) the basic reason — global warming — is stated, followed by two further points related to this reason, so this answer scored 3 marks.

(c) Explain the formation of **two fluvioglacial landforms.** [4 marks]

(d) Explain the importance of freeze–thaw weathering in the formation of periglacial landforms. [6 marks]

ⓔ Part (c) is a point-marked question. Take care to choose fluvio-glacial landforms not any glacial landforms. In addition 2 × 2 mark developed explanations are needed rather than a long explanation of only one landform. Part (d) is a Level-marked question (see Levels mark scheme on page 90). Answers need to demonstrate an understanding of frost-shattering and link it to landforms (plural). Questions such as this, which focus on physical processes, need good use of terminology and logical, sequential explanations.

Student answer

(c) Eskers are fluvioglacial landforms which consist of long narrow ridges of sediment. They formed by deposition from subglacial or englacial streams as velocity drops and sediment is deposited and the sediment has some rounding and sorting indicating transport by meltwater. Sandhurs are outwash plains at the snout of a glacier, and are formed by deposition as fast-flowing meltwater slows and sediment is reworked by meltwater channels flowing across the sandhur with continued deposition building up layers of sediment.

(d) Freeze–thaw weathering is one of the most important physical processes in periglacial areas. It relies on repeated freezing–thawing cycles which is a feature of the periglacial climate. Water freezes in cracks and expands by 9% in volume exerting a force that can split rocks into angular fragments. These angular fragments are the basis of patterned ground formation as the related process of frost-heave moves larger freeze–thaw weathered rocks towards the surface to form striped polygons. Frost-heave is the repeated expansion and contraction of the permafrost layer from season to season. Most periglacial landforms, such as solifluction lobes and terraces, and blockfields, consist of freeze–thaw debris.

ⓔ 10/10 marks awarded. The answer to Part (c) scores 4 marks. There is no mark for naming the landform or describing it. The esker explanation has two valid explanations: deposition in sub-glacial and englacial streams and the sediment type indicating formation process. The sandur explanation also has two explanations: deposition as meltwater slows and the idea of layers building up.

92 Edexcel Geography

Exam-style questions

Commentary on the questions

Tips on what you need to do to gain full marks, indicated by the icon **ⓔ**

Sample student answers

Practise the questions, then look at the student answers that follow.

Commentary on sample student answers

Read the comments (preceded by the icon **ⓔ**) showing how many marks each answer would be awarded in the exam and exactly where marks are gained or lost.

■About this book

Much of the knowledge and understanding needed for AS and A-level geography builds on what you have learned for GCSE geography, but with an added focus on key geographical concepts, depth of knowledge and understanding of content. This guide offers advice for the effective revision of physical geography, which all students need to complete.

The first part of the A-level Paper 1 tests your knowledge and application of aspects of physical geography with a particular focus on **Tectonic processes and hazards**, and *one* of **Glaciated landscapes and change** and **Coastal landscapes and change**. The whole exam (including the other areas of study not covered here) lasts 2 hours and 15 minutes and makes up 30% of the A-level qualification. The same topics and approach make up 50% of the AS Paper 1, which lasts 1 hour and 45 minutes. More information on the external exam papers is given in the Question & Answer section at the back of this book.

To be successful in this unit you have to understand:
■ the key ideas of the content
■ the nature of the assessment material — by reviewing and practising sample structured questions
■ how to achieve a high level of performance within the exams.

This guide has two sections:

Content Guidance — this summarises some of the key information that you need to know to be able to answer the exam questions with a high degree of accuracy and depth. In particular, the meaning of key terms is made clear and details of case study material are provided to help meet the spatial context requirement within the specification.

Questions & Answers — this includes sample questions similar in style to those you might expect in the exam. There are sample student responses to these questions as well as detailed analysis, which will give further guidance about what exam markers are looking for to award top marks.

The best way to use this book is to read through the relevant topic area first before practising the questions. Only refer to the answers and examiner comments after you have attempted the questions.

Content Guidance

This section outlines the following areas of the AS geography and A-level geography specifications:

- Tectonic processes and hazards
- Landscape systems, processes and change
  - Glaciated landscapes and change
  - Coastal landscapes and change

Read through the topic area before attempting a question from the Questions & Answers section.

■ Tectonic processes and hazards

Why are some locations more at risk from tectonic hazards?

- Tectonic hazards (earthquakes, volcanic eruptions and tsunami) occur in specific locations, related to tectonic plate boundaries and other tectonic settings.
- Their distribution is uneven, with some areas at high risk and other locations at no risk.
- Tectonic events can generate multiple hazards when they occur.

The global distribution of tectonic hazards

All tectonic hazards are caused by the Earth's internal heat engine. Radioactive decay of isotopes such as uranium-238 and thorium-232 in the Earth's core and mantle generate huge amounts of heat which flow towards the Earth's surface. This heat flow generates convection currents in the plastic mantle. The interior of the Earth is therefore dynamic rather than static. Most tectonic hazards occur at or near tectonic plate boundaries. These represent the locations of ascending (divergent plate boundaries) and descending (convergent plate boundaries) arms of mantle convection cells (Figure 1).

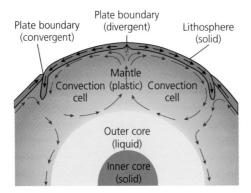

Figure 1 Earth's internal structure and mantle convection

Exam tip

All tectonic hazards are physical events, with natural causes. Only use the word 'disaster' when referring to the impact of these events on people.

The mantle is a solid, but because of the very high temperatures present it is deformable (plastic) and capable of very slow 'flow'.

Exam tip

The terms 'plate boundary' and 'plate margin' are used interchangeably. They both mean the narrow, linear zone where two tectonic plates meet.

Figure 2 shows the distribution of earthquakes, volcanoes and tectonic plate boundaries. It is clear that most earthquakes occur at, or close to, these boundaries. This is also true of volcanic eruptions. Some plate boundary earthquakes cause a secondary tectonic hazard, tsunami.

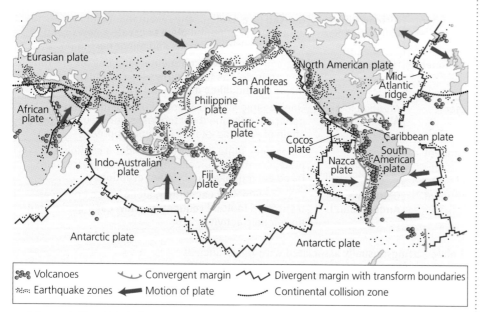

Figure 2 The global distribution of plate boundaries, earthquakes and volcanoes

Not all tectonic plate boundaries are the same and this has an important impact on the type and magnitude of tectonic hazards. Plate boundary type depends on two factors.

1 Motion: whether plates are moving apart (divergent), colliding (convergent) or sliding past each other (conservative or transform).

2 Plate type: whether the tectonic plates are oceanic or continental. Oceanic plates make up the ocean floor and are high density, basaltic rock but only 7–10 km thick. Continental plates make up the Earth's landmasses and are much thicker at 25–70 km but made of less dense, granitic rock.

This combination of motion and plate type yields a range of different plate boundary types that are summarised in Table 1.

Although the vast majority of earthquakes and eruptions occur at plate boundaries, a small number do not. Some volcanic eruptions are described as 'intra-plate'. This means they are distant from a plate boundary at locations called mid-plate hotspots (such as Hawaii and the Galapagos Islands). At these locations:

- isolated plumes of convecting heat, called **mantle plumes**, rise towards the surface, generating basaltic volcanoes that tend to erupt continually

- a mantle plume is stationary, but the tectonic plate above moves slowly over it

- over millennia, this produces a chain of volcanic islands, with extinct ones most distant from the plume location.

Mantle plumes are concentrated areas of heat convection. At plate boundaries they are sheet-like, whereas at hot spots they are column-like.

Table 1 Different plate boundary settings

Boundary type	Plate type	Example	Description and hazards
Divergent	Oceanic–Oceanic	Mid-Atlantic ridge at Iceland	Rising convection currents bring magma to the surface resulting in small, basaltic eruptions, creating new oceanic plate. Minor, shallow earthquakes
	Continent–Continent	African Rift Valley/Red Sea	Caused by a geologically recent mantle plume splitting a continental plate to create a new ocean basin. Basaltic volcanoes and minor earthquakes
Convergent	Continent–Continent	Himalayas	The collision of two continental landmasses creating a mountain belt as the landmasses crumple. Infrequent major earthquakes distributed over a wide area
	Oceanic–Oceanic	Aleutian Islands, Alaska	One oceanic plate is subducted beneath another, generating frequent earthquakes and a curving (arc) chain of volcanic islands (violent eruptions)
	Oceanic–Continent	Andean Mountains	An oceanic plate is subducted under a continental plate, creating a volcanic mountain range, frequent large earthquakes and violent eruptions
Conservative	Oceanic–Continent	California, San Andreas fault zone	Plates slide past each other, along zones known as transform faults. Frequent, shallow earthquakes but no volcanic activity

Earthquakes can occur in mid-plate settings, usually associated with major ancient fault lines being re-activated by tectonic stresses. For instance, the New Madrid Seismic Zone on the Mississippi River generates earthquakes up to magnitude 7.5 but is thousands of miles from the nearest plate boundary.

Theories of plate motion

Earth's tectonic plates move at a speed of 2–5 cm per year. There are seven very large major plates (African, Pacific), smaller minor plates (Nazca, Philippine Sea) and dozens of small microplates. All fit together into a constantly moving jigsaw of rigid lithosphere. Each plate is about 100 km thick. Its lower part consists of upper mantle material while its upper part is either oceanic or continental crust.

The theory of plate tectonics has developed because of a number of key discoveries:

- Alfred Wegener's Continental Drift hypothesis in 1912 that postulated that now-separate continents had once been joined
- the ideas of Arthur Holmes in the 1930s that Earth's internal radioactive heat was the driving force of mantle convection that could move tectonic plates
- the discovery in 1960 of the asthenosphere, a weak, deformable layer beneath the rigid lithosphere on which the lithosphere moves
- the discovery in the 1960s of magnetic stripes in the oceanic crust of the sea bed; these are palaeomagnetic (ancient magnetism) signals from past reversals of Earth's magnetic field and prove that new oceanic crust is created by the process of sea-floor spreading at mid-ocean ridges
- the recognition of transform faults by Tuzo Wilson in 1965.

It remains a theory because scientists have not yet directly observed the interior of the Earth.

Knowledge check 2

Why is plate tectonics still a theory rather than proven fact?

Constructive margins

Mantle convection forces plates apart at constructive plate margins. Tensional forces open cracks and faults between the two plates. These create pathways for magma to move towards the surface and erupt, creating new oceanic plate. Eruptions are small and effusive in character, as the erupted basalt lava has a low gas content and high viscosity. Earthquakes are shallow, less than 60 km deep, and have low magnitudes of under 5.0.

Destructive margins and subduction zones

Locations where one plate is subducted beneath another illustrate the forces that drive plate tectonics (Figure 3).

- Mantle convection pulls oceanic plates apart, creating the fracture zones at constructive margins, and convection also pulls plates towards subduction zones.
- Constructive margins have elevated altitudes because of the rising heat beneath them, which creates a 'slope' down which oceanic plates slide (gravitational sliding).
- Cold, dense oceanic plate is subducted beneath less dense continental plate; the density of the oceanic plate pulls itself into the mantle (slab pull).

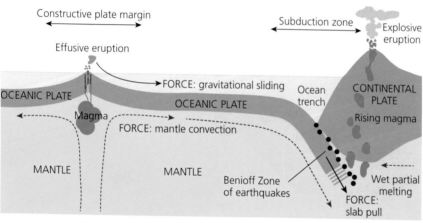

Figure 3 Tectonic forces and plate margin features

Earthquakes at **subduction** zones occur at a range of focal depths from 10 km to 400 km, following the line of the subducting plate. This is called a Benioff Zone, and it can yield very large earthquakes up to magnitude 9.0. The descending plate begins to melt at depth by a process called wet partial melting. This generates magma with a high gas and silica content, which erupts with explosive force.

Collision zones

The Himalaya mountains are a location where two continental plates are in collision (the Indo-Australian and Eurasian plates). The collision began about 52 million years ago. As both continental plates have the same low density, subduction is not possible. Instead, the plates have 'crumpled', creating enormous tectonic uplift in the form of the Himalaya and Tibetan Plateau. Magma is being generated at depth, but it cools and solidifies beneath the surface so eruptions are very rare. Collision zones are cut by huge thrust faults that generate shallow, high-magnitude earthquakes such as in Kashmir in 2005 and Nepal in 2015.

Subduction is the process of one plate sinking beneath another at a convergent plate boundary.

> **Exam tip**
>
> Sometimes it can be quicker and easier to sketch and annotate a diagram of a plate boundary in the exam.

Transform zones

Conservative plate boundaries consist of transform faults. These faults 'join up' sections of constructive plate boundary as they traverse the Earth's surface in a zig-zag pattern. In some locations, long transform faults act like a boundary in their own right, most famously in California where a fault zone — including the San Andreas fault — creates an area of frequent earthquake activity. Earthquakes along conservative boundaries often have shallow focal depths, meaning high-magnitude earthquakes can be very destructive. Volcanic activity is absent.

The causes of tectonic hazards

Earthquakes

Earthquakes are a sudden release of stored energy. As tectonic plates attempt to move past each other along fault lines, they inevitably 'stick'. This allows strain to build up over time and the plates are placed under increasing stress. Earthquakes are generated because of sudden release of the stress — so-called 'stick-slip' behaviour. A pulse of energy radiates out in all directions from the earthquake **focus** (point of origin) (Figure 4). In some cases the earthquake motion displaces the surface, so a fault scarp can be seen.

An earthquake originates at the **focus**. The epicentre is the point on the Earth's surface directly above the focus.

Knowledge check 3

What is an earthquake?

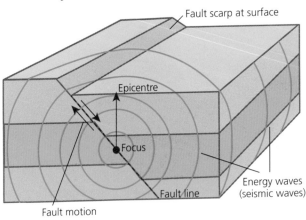

Figure 4 The anatomy of an earthquake

Earthquakes generate three types of seismic wave:

- P-waves, or primary waves, are the fastest. They arrive first, and cause the least damage
- S-waves, or secondary waves, arrive next and shake the ground violently, causing damage
- L-waves, or Love waves, arrive last as they travel only across the surface. However, they have a large amplitude and cause significant damage, including fracturing the ground surface.

Earthquakes cause crustal fracturing within the Earth, but also buckle and fracture the ground surface. Some very large earthquakes, such as the one that generated the 2004 Indian Ocean tsunami, rupture a fault line for up to 1000 km. Think of this as 'unzipping' a fault, with energy pulses being generated along the entire fault length. Such earthquakes can cause ground shaking that lasts up to 5 minutes, as well as dozens of powerful aftershocks.

Earthquakes frequently generate large landslides as secondary hazards. This is especially true in areas of geologically young (and therefore unstable) mountains such as the Himalaya. Landslides accounted for up to 30% of deaths in the 2008 Sichuan and 2005 Kashmir earthquakes.

Liquefaction is a particular hazard in areas where the ground consists of loose sediment such as silt, sand or gravel that is also waterlogged — often found in areas close to the sea or lakes. Intense earthquake shaking compacts the loose sediment together, forcing water between the sediment out and upward. This undermines foundations and causes buildings to sink, tilt and often collapse.

Volcanoes

Major volcanic eruptions frequently have more than one hazard associated with them. In some cases these are secondary hazards, which are an indirect consequence of the eruption (lahar, jökulhlaup). This is especially the case with the violent eruptions associated with volcanoes at destructive plate margins, as shown in Table 2.

Table 2 Volcanic hazards

Volcanic hazard	Explanation	Volcano types
Lava flow	Extensive areas of solidified lava, which can extend several kilometres from volcanic vents if the lava is basaltic and low viscosity. It can flow at up to 40 kmh	✚ ●
Pyroclastic flow	Very large, dense clouds of hot ash and gas at temperatures of up to 600°C. They can flow down the flanks of volcanoes and devastate large areas	✚
Ash fall	Ash particles, and larger tephra particles, can blanket huge areas in ash, killing vegetation, collapsing buildings and poisoning water courses	■ ✚
Gas eruption	The eruption of carbon dioxide and sulphur dioxide, which can poison people and animals in extreme cases	✚ ●
Lahar	Volcanic mudflows, which occur when rainfall mobilises volcanic ash. They travel at high speed down river systems and cause major destruction	✚
Jökulhlaup	Devastating floods caused when volcanoes erupt beneath glaciers and ice caps, creating huge volumes of meltwater. They are common in Iceland	■

✚ Subduction zone volcano (composite type)

● Hot-spot volcano (shield type)

■ Constructive plate margin volcano (cinder cone, fissure eruption)

In most cases, only large composite volcanoes found at destructive plate margins represent a significant tectonic hazard. These eruptions often have lava flows, pyroclastic flows, lahars and extensive ash and tephra fall that can affect areas up to 30 km from the volcanic vent.

Knowledge check 4

Which type of plate boundary produces the most hazardous volcanoes?

Pyroclasts (meaning 'fire broken') are any rock fragments ejected from a volcano including ash, tephra and volcanic bombs.

Tsunami

There have been a number of deadly tsunami in recent years (Table 3) such that this hazard is more well-known than it once was. Tsunami can be generated by landslides and even eruptions of volcanic islands. Most are generated by sub-marine earthquakes at subduction zones.

Table 3 Deadly tsunami since 2004

Location	Earthquake magnitude	Wave height (m)	Deaths
2004 Indian Ocean	9.2	24	230,000
2006 Java, Indonesia	7.7	2–6	800
2009 Samoa	8.1	14	190
2011 Japan	9.0	9.3	16,000

Exam tip

Tsunami have nothing to do with tides, despite being commonly called 'tidal waves'.

Tsunami are generated when a sub-marine earthquake displaces the sea bed vertically (either up or down) as a result of movement along a fault line at a subduction zone. The violent motion displaces a large volume of water in the ocean water column, which then moves outward in all directions from the point of displacement. The water moves as a vast 'bulge' in open water, rather than as a distinct wave. Tsunami characteristics are very different from those of wind-generated ocean waves:

- wave heights are typically less than 1 m
- wavelengths are usually more than 100 km
- speeds are 500–950 kmh.

Knowledge check 5

What has to happen to the sea bed during an earthquake for a tsunami to be generated?

In the open ocean tsunami waves are barely noticeable. As the waves approach shore they slow dramatically, wavelength drops but wave height increases. Tsunami usually hit coastlines as a series of waves (a 'wave-train') the effect of which is more akin to a flood than a breaking wave. Sub-marine earthquakes that occur close to shorelines can generate intense ground-shaking damage, followed by damage from the subsequent tsunami.

Why do some tectonic hazards develop into disasters?

- There is a relationship between hazard, vulnerability and resilience that sometimes leads to tectonic disasters.
- The impacts of disasters vary hugely between locations.
- Tectonic disasters and their impacts can be understood using hazard profiles.
- Level of development has a major influence on the nature of disasters.
- Governance is a key concept in understanding the impacts of disasters.

Exam tip

Use key words from the specification in your answers. Although related, the word 'governance' has a different meaning to 'management'.

Hazard versus disaster

Tectonic hazards are natural events that have the potential to harm people and their property. A disaster is the realisation of a hazard, i.e. harm has occurred. By definition disasters have to involve people and they occur at the intersection of people and hazards as shown by the Degg disaster model (Figure 5).

A **threshold** level is often used to determine whether the impact of an event is large enough to be considered a disaster, such as:

- 10 or more deaths

A **threshold** is the magnitude above which a disaster occurs. This threshold level could be different in a developed versus a developing country because of different levels of resilience.

- 100 or more people affected
- US$1 million in economic losses.

We can understand the relationship between hazards and disasters using the hazard risk equation. The risk of disaster rises if hazard magnitude rises, i.e. a VEI 6 eruption compared with a VEI 3. If vulnerability rises (poverty, lack of preparedness, lack of awareness of potential hazards) so does risk:

$$risk = \frac{hazard \times vulnerability}{capacity\ to\ cope}$$

Some communities have a high capacity to cope and high **resilience**. This means they can reduce the chances of disasters occurring because:

- they have emergency evacuation, rescue and relief systems in place
- they react by helping each other to reduce numbers affected
- hazard-resistant design or land-use planning have reduced the numbers at risk.

For these communities the threshold for a disaster will be higher than for ones with low coping capacity.

Some disasters are truly catastrophic in terms of their impact. A good example is the 2010 Port-au-Prince earthquake in Haiti. Its magnitude of 7.0 was relatively low, but the death toll has been estimated at 160,000. The Pressure and Release model (PAR model) can be used to help understand the enormous death toll (Figure 6).

Progression of vulnerability				Natural hazards
→ **Root causes**	**→** **Dynamic pressures**	**→** **Unsafe conditions**	Disaster (risk = hazard × vulnerability)	**←** Earthquake
Low access to resources	Lack of education, training and investment	Poor construction standards		Eruption
Limited influence in decision making		Unsafe infrastructure		Tsunami
	Rapid population change and urbanisation			
Poor governance and a weak economic system		Poverty		
		Lack of social safety net		

Figure 6 The Pressure and Release Model
Source: after Blaikie (1994)

The PAR model suggests that the socio-economic context of a hazard is important. In poor, badly governed (root causes) places with rapid change and low capacity (dynamic pressures) and low coping capacity (unsafe conditions) disasters are likely. Sadly, in 2010 Haiti fitted many of these criteria (Table 4).

Table 4 The PAR model applied to the 2010 Haiti earthquake

Per capita GDP (PPP) US$1200	70% of jobs are in the informal sector	25% of people live in extreme poverty
50% of the population is under 20 years old	Port-au-Prince, a city of 3.5 million, has no sewer system	80% of Port-au-Prince's housing is unplanned, informal slums

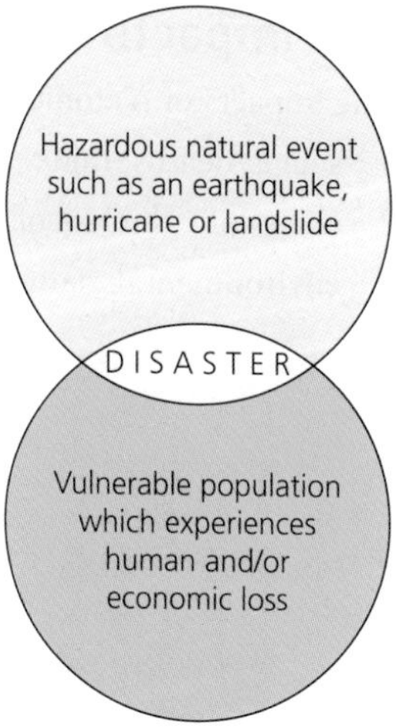

Figure 5 The Degg disaster model

Resilience is the ability of a community to cope with a hazard; some communities are better prepared than others so a hazard is less likely to become a disaster.

Exam tip

You can draw models, such as the Degg disaster model, in your answers to exam questions.

Knowledge check 6

What is the name given to the outcome between a vulnerable population and a natural hazard?

The impacts of tectonic hazards

The impacts of tectonic hazards are broadly of three types:

1 social: deaths, injury and wider health impacts including psychological ones
2 economic: the loss of property, businesses, infrastructure and opportunity
3 environmental: damage or destruction of physical systems, especially ecosystems.

In the last 30 years, different tectonic hazards have had contrasting impacts in terms of scale (Table 5).

Comparing impacts between countries is difficult because both the physical nature of the event and the socio-economic profiles of affected places are different. Table 7 on page 15 compares the impact of recent disasters in developed, developing and emerging countries. Some general observations are:

- economic costs in developed and emerging economies are, in some cases, enormous
- deaths in developed countries are low, except for the 2011 Japanese tsunami (a very rare megadisaster)
- volcanic eruption impacts are small compared with those of earthquakes and tsunami.

Measuring magnitude and intensity

The 'size' of a tectonic event, called its magnitude, has a relationship with its impact. Broadly, larger magnitude events have a bigger impact — but the relationship is not a simple one because of the vulnerability and capacity to cope parts of the hazard risk equation.

Earthquake magnitude is measured using the moment magnitude scale (MMS). This is an updated version of the well-known Richter magnitude scale. MMS measures the energy released during an earthquake. This is related to the amount of slip (movement) on a fault plane and the area of movement on the fault plane. MMS uses a logarithmic scale, meaning that a magnitude 6 earthquake has ten times more ground shaking than a magnitude 5.

The Mercalli scale measures earthquake intensity on a scale of I–XII. This older scale measures what people actually feel during an earthquake, i.e. the intensity of the shaking effects not the energy released. It cannot be easily used to compare earthquakes as shaking experienced depends on building type and quality, ground conditions and other factors.

The relationship between magnitude and death toll is a weak one because:

- some earthquakes cause serious secondary impacts, such as landslides and tsunami
- earthquakes hitting urban areas have greater impacts than those in rural areas
- level of development, and level of preparedness, affect death tolls
- isolated, hard to reach places could have a higher death toll because rescue and relief take longer.

The magnitude of a volcanic eruption is measured using the volcanic explosivity index (VEI). VEI ranges from 0 to 8 and is a composite index combining eruption height, volume of material (ash, gas, tephra) erupted and duration of eruption (Table 6).

Table 5 Contrasting tectonic hazard impacts

Volcanic eruptions
Small and declining impacts, especially death tolls

Earthquakes
Large impacts, as significant earthquakes are common and widespread

Tsunami
Very large impacts from a small number of events

A megadisaster is a disaster with unusually high impacts. Today that means millions of people affected and billions of dollars in damage over a wide area, i.e. an entire region or even more than one country.

Knowledge check 7

What type of scales are the Richter and moment magnitude scales?

Exam tip

Learn some key facts and figures about your examples and case studies as these add weight to your answers.

Exam tip

Make sure you understand the difference between magnitude and intensity.

Table 6 The Volcanic Explosivity Index scale

VEI	0	1	2	3	4	5	6		8
Eruption height	<100 m	100 m–1 km	1–5 km	3–15 km	>10 km	>10 km	>20 km		>20 km
Eruption volume	<10,000 m^3	>10,000 m^3	>0.001 km^3	>0.01 km^3	>0.1 km^3	>1 km^3	>10 km^3		>1000 km^3
	Effusive		Explosive					Colossal	

VEI eruptions from 0 to 3 are associated with shield volcanoes and basaltic eruptions at constructive plate margins and mid-plate hotspots. VEI eruptions from 4 to 7 occur at destructive plate margins, erupting high viscosity, high gas, high silica andesitic magma. No modern human has experienced a VEI 8 **supervolcano**. These are rare caldera eruptions such as Yellowstone and Toba.

A **supervolcano** is one whose impacts would be felt globally, because of a worldwide cooling of the Earth's climate, perhaps for up to 5 years.

Table 7 The impact of disasters compared

	Developed countries	Emerging countries	Developing countries
Volcanic eruption	2010 Eyjafjallajökull (Iceland) Constructive margin, mid-ocean ridge Basaltic magma; stratovolcano VEI = 4	2010 Merapi (Indonesia) Destructive margin subduction zone Andesitic magma; composite cone volcano VEI = 4	2002 Nyiragongo (DRC) Constructive margin, continental rift zone Basaltic magma, stratovolcano VEI = 1
	■ No injuries, no deaths ■ Major disruption to European and transatlantic air travel affecting 10 million passengers and costing US$1.7 billion in economic losses ■ Ice melt on the volcano caused some flash flooding	■ 353 deaths, about 500 injured ■ 350,000 people successfully evacuated before the eruption ■ US$0.6 billion in losses ■ Loss of rice harvest due to ash fall and some destruction of forests due to pyroclastic flows	■ 147 deaths, 120,000 made homeless ■ 15% of the city of Goma destroyed by lava flows ■ Major international aid response launched ■ US$1.2 billion in economic losses
Earthquake	2010 Canterbury (New Zealand) Magnitude 7.1 Focal depth = 10 km Subduction zone	2008 Sichuan (China) Magnitude 8.0 Focal depth = 19 km Continent–Continent collision zone	2015 Gorkha (Nepal) Magnitude 7.9 Focal depth = 8.2 km Continent–Continent collision zone
	■ 100 injuries, no deaths ■ Widespread building damage due to **liquefaction** ■ A magnitude 6.3 aftershock in 2011 in Christchurch killed 185 ■ Total costs (including aftershocks) estimated at US$40 billion	■ 69,000 deaths, 370,000 injured ■ At least 5 million homeless ■ More than US$140 billion in economic losses ■ About one-third of deaths due to landslides	■ 9000 deaths, 22,000 injured ■ Economic losses about US$5 billion ■ Killer avalanches triggered on Mt Everest ■ Many rural villages totally destroyed
Tsunami	2011 Tohoku (Japan) Magnitude 9.0 9.3 m tsunami height Megathrust subduction zone	2010 Talcahuano (Chile) Magnitude 8.8 2.6 m tsunami height Megathrust subduction zone	2004 Indian Ocean Magnitude 9.2 24 m tsunami height Megathrust subduction zone
	■ 16,000 deaths and 6000 injuries ■ US$300 billion in economic losses ■ 46,000 buildings destroyed and 145,000 damaged ■ Huge infrastructure damage to ports, water and electricity supply	■ 525 deaths ■ 370,000 homes damaged ■ Economic costs of US$15–30 billion ■ Overall, the earthquake was more damaging than the accompanying tsunami	■ 230,000 deaths and 125,000 injured ■ 1.7 million people displaced across 15 countries ■ US$15 billion in economic losses

Content Guidance

Hazard profiles

Tectonic events can be compared using hazard profiles. These allow a better understanding of the nature of hazards, and therefore risks associated with each. Figure 7 shows hazard profiles for four tectonic hazards. Hazards with the following characteristics present the highest risk:

- high magnitude, low frequency events — these are the least 'expected' as, by definition, they are unlikely to have occurred in living memory
- rapid onset events with low spatial predictability — they could occur in numerous places, and happen without warning
- regional areal extent — affecting large numbers of people in a wide range of locations.

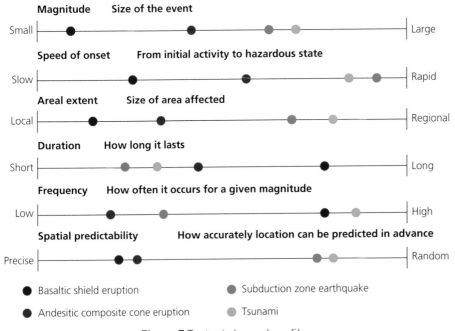

- Basaltic shield eruption
- Andesitic composite cone eruption
- Subduction zone earthquake
- Tsunami

Figure 7 Tectonic hazard profiles

Arguably, major earthquakes at subduction zones and collision zones are the most dangerous tectonic hazards. They can have magnitudes of 8–9 MMS, cannot be predicted and could occur along any of tens of thousands of kilometres of plate margin, instantaneously.

The 2005 Kashmir earthquake ranks as one of the most destructive in recent decades.

- At magnitude 7.6 and with a ground shaking intensity of VII (severe) this was a large event.
- As with all earthquakes, speed of onset was very rapid so there was no chance of evacuating to a safe area.
- Damage was centred on Muzaffarabad but spread over an areal extent of more than 1000 km².
- Ground shaking lasted 30–45 seconds, but landslides triggered by the earthquake continued for some time, as did aftershocks up to magnitude 6.4.

Destruction in Kashmir was severe: 87,000 deaths, 2.8 million people displaced or made homeless, 17,000 schools destroyed or damaged and nearly 800 health centres

Liquefaction occurs in waterlogged, loose sediment; earthquake shaking 'liquefies' the ground, causing buildings to tilt, sink and collapse.

Knowledge check 8

Which plate tectonic setting generates the most destructive, high-risk earthquakes?

destroyed. Numerous factors including poverty, poor building construction, time of day (many children were in school), geology and terrain (contributing to landslides) and isolation (making the rescue and relief effort difficult) help explain the impacts, but two hazard profile characteristics are also relevant:

1 frequency: the previous major earthquake in Kashmir was in 1905, so there was no 'collective memory' of the risks and impacts of earthquakes in the region

2 spatial predictability: Kashmir is in a 'seismic gap', i.e. an area of known risk that had not experienced an earthquake for some time. This should have been acted upon through education and risk awareness, which could have reduced the impacts.

Development and governance

The impacts of tectonic hazards are not only determined by level of development, but it is an important factor. Table 8 shows five recent earthquakes which all had the same 7.7 magnitude. There is some relationship between death toll and HDI (Human Development Index), such that lower HDI appears to suggest higher death tolls.

Table 8 Five magnitude 7.0 earthquakes compared

Date and location	Deaths	HDI
2013 Balochistan, Pakistan	825	0.54
2015 Nepal	9018	0.55
2010 Sumatra, Indonesia	711	0.68
2013 Khash, Iran	35	0.77
2007 Chile	2	0.83

However, other factors such as population density, duration of ground shaking, secondary hazards and response are also important. Generally, low level of development increases risk by increasing vulnerability, as shown in Table 9.

Table 9 Factors increasing and mitigating risk

Increasing risk	Mitigating risk
■ Population growth ■ Urbanisation and urban sprawl ■ Environmental degradation ■ Loss of community memory about hazards ■ Very young, or very old, population ■ Ageing, inadequate infrastructure ■ Greater reliance on power, water, communication systems	■ Warning and emergency-response systems ■ Economic wealth ■ Government disaster-assistance programmes ■ Insurance ■ Community initiatives ■ Scientific understanding ■ Hazard engineering

In some locations with very low levels of human development (HDI below 0.55) vulnerability is usually high because:

■ many people lack basic needs of sufficient water and food even in 'normal' times
■ much housing is informally constructed with no regard for hazard resilience
■ access to healthcare is poor, and disease and illness are common
■ education levels are lower, so hazard perception and risk awareness is low.

Many low-income groups lack a 'safety net' — either a personal one (savings, food stores) or a government one (social security, aid, free healthcare) — so have few resources after a disaster.

Aftershocks occur in the hours, days and months after a primary earthquake and can be of high magnitude; they often number in the hundreds or thousands.

Exam tip

Physical process exam questions require you to use accurate and precise terminology.

In rural Nepal, the area hit by the 2015 earthquake, 40% of families live below the poverty line and more than 90% of people depend on subsistence farming. Of the rural population, 40% exhibit stunting as a result of malnutrition and only 20–40% of rural adults are literate.

Governance

Governance refers to the processes by which a country or region is run. Sometimes this is called 'public administration' and relates to how 'well-run' a place is. 'Good governance' implies that national and local government are effective in keeping people safe, healthy and educated.

The effectiveness of governance varies enormously and has a significant impact on coping capacity and resilience in the event of a natural disaster. Table 10 explains the link between governance and vulnerability.

Table 10 Aspects of governance and disaster vulnerability

Meeting basic needs	Planning	Environmental management
When food supply, water supply and health needs are met the population is physically more able to cope with disaster	Land-use planning can reduce risk by preventing habitation on high risk slopes, areas prone to liquefaction or areas within a volcanic hazard risk zone	Secondary hazards, such as landslides, can be made worse by deforestation. The right monitoring equipment can warn of some hazards, such as lahars
Preparedness	**Corruption**	**Open-ness**
Education and community preparation programmes raise awareness and teach people how to prepare, evacuate and act	Siphoning off money ear-marked for hazard management, or 'kick-backs' and bribes to allow illegal or unsafe buildings increase vulnerability	Governments that are open, with a free press and media, can be held to account, increasing the likelihood that preparation and planning take place

Most countries have national disaster management agencies, such as FEMA in the USA, which increase resilience to hazards and reduce the impacts of disasters. In the developing world these can be effective, such as PHIVOLCS in the Philippines, but they are often under-funded and under-resourced. Low-level corruption of local government officials is common in many developing countries, meaning that building codes are often ignored and construction allowed in inappropriate places. This was widely blamed for the high death toll of 17,000 in the Izmit earthquake in Turkey in 1999.

Geographical factors

The nature of tectonic hazard impacts is influenced by a number of geographical factors, these include:

- population density: highly populated areas may be hard to evacuate, such as the area around Mt Vesuvius in Italy, and are likely to be hit harder by an earthquake
- degree of urbanisation: when cities are struck by major earthquakes, such as the 1995 Kobe earthquake in Japan or Haiti in 2010, death tolls can be high because of the concentration of at-risk people
- isolation and accessibility: often rural areas are hit less hard than urban areas by the initial impact of a tectonic disaster, but isolation and limited access can slow the rescue relief effort. The 2005 Kashmir earthquake is a good example.

Urban areas usually have more assets than rural areas. These include hospitals, emergency services, food stores and transport connections, which increase resilience and coping capacity compared with isolated rural places. However, high population density may mean more people are affected.

Knowledge check 9

Name a secondary hazard often associated with earthquakes in high terrain areas.

Corruption refers to illegal practices, such as accepting bribes designed to influence decision making or paying people to stay silent about known problems.

Knowledge check 10

Why are cities more vulnerable to earthquakes than rural areas?

Disaster context

No tectonic disaster can be separated from the wider local and national context within which it occurs. Table 11 illustrates this (see Table 7 on page 15 for additional examples).

Table 11 Disaster context in Haiti and China compared

Developing country Haiti HDI = 0.48	Emerging country China HDI = 0.73
Port-au-Prince earthquake, 2010 160,000 deaths, 1.5 million homeless, 250,000 homes destroyed	Sichuan earthquake, 2008 69,000 deaths, 375,000 injured and economic costs of US$140 billion
Decades of corrupt, ineffective and brutal government left Haitian people hugely vulnerable because of slum housing, ineffectual water supply and endemic poverty. A post-earthquake cholera epidemic had infected 700,000 and killed 9000 by 2015	Economic losses in China were high, reflecting its development progress since 1990 (destroyed formal homes, businesses and infrastructure). The immediate response was rapid because the 2008 Beijing Olympic Games were only months away, so the Communist government mobilised the army and other responders rapidly

In developed countries major death tolls from tectonic hazards are rare. The 2011 Tohoku earthquake and tsunami in Japan is very much exceptional in terms of impacts. Countries such as Japan, the USA and Chile have:

- advanced and widespread insurance, allowing people to recover from disasters — at least in the long term
- government-run preparations such as Japan's Disaster Prevention Day on 1 September each year, as well as public education about risk, coping, response and evacuation
- sophisticated monitoring of volcanoes and, where possible, defences such as tsunami walls
- regulated local planning systems, which use land-use zoning and building codes to ensure buildings can withstand hazards and are not located in areas of unacceptable risk.

Land-use zoning is a planning tool used to decide what type of buildings (residential, commercial, industrial or none) are allowed in particular locations.

How successful is the management of tectonic hazards and disasters?

- There are important trends in tectonic disasters which need to be understood.
- Some locations are vulnerable to multiple hazards.
- Prediction and forecasting are important in managing disasters, but are not always possible.
- There are different models of disaster management but they are not universally applied.
- Different strategies can modify hazard events, vulnerability and loss, and reduce the impacts of events.

Tectonic disaster trends

The number of disasters and the impacts of disasters are not static. There is, however, a difference between two broad categories of natural hazard.

1 Hydro-met hazards, such as floods, storms, cyclones and drought, appear to have become more common over time, perhaps because of global warming and human environmental management issues such as deforestation.

2   Tectonic hazards, i.e. the events, have not increased or decreased over time. The number of events is broadly the same decade on decade.

Tectonic hazards and disasters are not the same, so even though the number of hazard events remains stable the number of disasters has risen. Trends for all disaster types are shown in Figure 8.

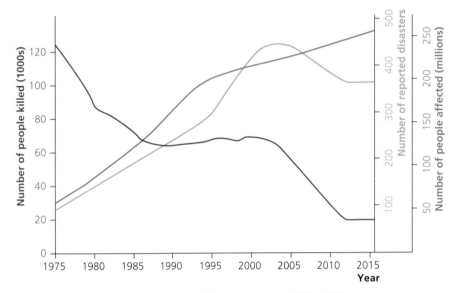

Figure 8 Natural disaster trends 1975–2015

Figure 8 shows three trends for all disasters.

1   Deaths have fallen over time because of better response management, preparation and, in some cases, prediction. Numbers of deaths have fallen especially since 2000, which may be due to vastly improved mobile communications to warn people of disasters.

2   The number of reported disasters increased then stabilised as improvements in data coverage and the accuracy of databases increased. Decades ago many disasters in isolated areas simply went unreported. Most recently, numbers of reported disasters have fallen, suggesting fewer hazard events are becoming disasters.

3   The number of people affected by disasters continues to rise as populations grow and more people live in risky locations.

Trends in tectonic hazards can be summarised as follows.

■ There has been no change in the number of earthquake disasters since 1980, which varies between about 15 and 40 each year.
■ Earthquake deaths are very variable: there were fewer than 1000 deaths worldwide in 2012 and 2014, yet more than 200,000 in 2010 and 2004. Overall, there are fewer earthquake deaths than there were 30–40 years ago but the impact of single, mega-disasters skews the data.
■ The trend for earthquake economic losses is upwards, averaging about US$20–40 billion per year but, once again, this is affected by a few very large events.

Exam tip

You can sketch graphs, such as Figure 8, as part of your exam answer.

Mega-disasters are high magnitude, high impact, infrequent disasters that affect multiple countries directly or indirectly so their impacts are regional or even global.

Economic losses from tectonic disasters continue to rise. More people, who are more affluent, have more property to lose. This is increasingly true in emerging countries as well as developed ones.

Volcanic disasters are much less frequent than earthquake ones and deaths from eruptions are now rare. The last time an eruption killed more than 1000 people was in Cameroon in 1986 (Lake Nyos) and only seven eruptions since 1980 have killed more than 100 people. However, numbers affected can still be very large because of the mass evacuation of people around an erupting volcano, e.g. 130,000 affected by the eruption of Mt Merapi in Indonesia in 2010 but only 300 deaths.

Mega-disasters and multiple hazard zones

Very large tectonic disasters account for most deaths. There were about 270 deadly earthquakes between 2005 and 2015. Of the 433,000 people killed, 412,000 were killed by just five disasters. Three of these — Kashmir 2005, Sichuan 2008 and Nepal 2015 — are in the same tectonic location, i.e. the Himalaya collision zone. These three disasters account for 40% of all earthquake deaths between 2005 and 2015, and the 2010 Haiti earthquake accounts for another 50%.

In recent years three examples of what might be called 'mega-disasters' have occurred (Table 12). Although rare, these are characterised by impacts extending beyond the country immediately affected. The 2011 tsunami in Japan showed how the globalised, inter-dependent world economy could be affected by the economic and human impacts of disasters. In addition, the accompanying nuclear meltdown disaster at Fukushima was a catalyst in Germany abandoning its nuclear electricity programme.

Table 12 Mega-disaster impacts

	Number of countries affected	Impacts
2004 Asian tsunami	14 countries surrounding the Indian Ocean	Economic losses and deaths in Indonesia, Thailand, Sri Lanka and Somalia among others make this disaster one of the largest ever in terms of areal extent
2011 Japanese tsunami	Only Japan directly, but the economic impacts had global consequences	Disruption to ports, factories and power supplies had impacts for the global car-production supply chain and those of Boeing jets and semiconductors used in modern electronics
2010 Eyafjallajokul eruption	Over 20 European countries were affected by total or partial closure of their airspace	The ash cloud from the Eyafjallajokul eruption had a disruptive effect on air travel because of the dangers of jet engines ingesting ash: over 100,000 cancelled flights costing over £1 billion in losses

Exam tip

Learn some key facts and figures about named disaster events — they will make your answers more convincing.

A number of locations are **multiple hazard zones** (Figure 9). These include California, the Philippines, Indonesia and Japan. These locations:

■ are tectonically active and so earthquakes (and often eruptions) are common
■ are geologically young with unstable mountain zones prone to landslides

Multiple hazard zones are places where two or more natural hazards occur, and in some cases can interact to produce complex disasters.

- are often on major storm tracks either in the mid-latitudes or on tropical cyclone tracks
- may suffer from global climate perturbations such as El Niño/La Niña.

Famously, during the 1991 eruption of Mount Pinatubo in the Philippines the area was struck by Typhoon Yunga. Heavy rainfall from the typhoon mobilised volcanic ash into destructive lahars. This shows how linked hydrometeorological hazards can contribute to tectonic disasters. This eruption could have been significantly worse in terms of impact but it was successfully predicted and evacuation limited the death toll to about 850. In many earthquake-prone areas, landslides can be triggered by heavy rain on slopes previously weakened by earthquake tremors.

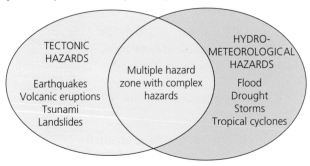

Figure 9 Multiple hazard zones

Prediction and forecasting

Prediction is the holy grail of hazard management but it is not always possible, as is outlined in Table 13. Despite decades of scientific research earthquakes cannot be predicted. However, the minimal death toll from volcanic eruptions (despite 60–80 eruptions worldwide per year) can be mainly attributed to vastly improved prediction of these events.

Prediction means knowing when, and where, a natural hazard will strike on a spatial and temporal scale that can be acted on meaningfully in terms of evacuation.

Forecasting is much less precise than predicting, and provides a 'percentage chance' of a hazard occurring, e.g. a 25% chance of a magnitude 7.0 earthquake in the next 20 years.

Table 13 Predicting tectonic hazards

Hazard type	Prediction?	Further details
Earthquakes	No	■ Only areas at high risk can be identified (risk **forecasting**), plus areas that are likely to suffer severe ground shaking and liquefaction; this can be used for land-use zoning purposes ■ 'Seismic gaps', i.e. areas that have not experienced an earthquake for some time and are 'overdue', can point to areas of especially high risk
Volcanic eruptions	Yes	■ Sophisticated monitoring equipment on volcanoes can measure changes as magma chambers fill and eruption nears ■ Tiltmeters and strain meters record volcanoes 'bulging' as magma rises and seismometers record minor earthquakes indicating magma movement ■ Gas spectrometers analyse gas emissions which can point to increased eruption likelihood
Tsunami	Partly	■ An earthquake-induced tsunami cannot be predicted ■ However, seismometers can tell an earthquake has occurred and locate it, then ocean monitoring equipment can detect tsunami in the open sea ■ This information can be relayed to coastal areas, which can be evacuated

Prediction of tsunami and eruptions depends on technology which has to be in place, operational and linked to warning dissemination and evacuation systems. Tsunami monitoring equipment was not present in the Indian Ocean in 2004 so there was no way of warning people on distant coasts — despite there being many hours in which to do so.

In many developing countries, volcano monitoring and tsunami warning may not be as good as they could be because of the cost of the technology. Also, it may be more difficult to reach isolated, rural locations with effective warnings.

Exam tip

Make sure you use the words 'prediction' and 'forecasting' carefully as they have very different meanings.

Hazard management

Prediction, when possible, is a vital part of attempts to manage the impacts of natural disasters. However, it is not the only approach. The hazard management cycle shown in Figure 10 (sometimes called the disaster management cycle) illustrates the different stages of managing hazards in an attempt to reduce the scale of a disaster. It is important to see this as a cycle, with one disaster event informing preparation for the next.

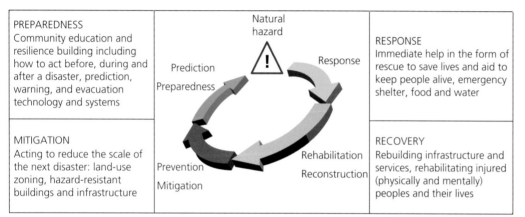

PREPAREDNESS
Community education and resilience building including how to act before, during and after a disaster, prediction, warning, and evacuation technology and systems

RESPONSE
Immediate help in the form of rescue to save lives and aid to keep people alive, emergency shelter, food and water

MITIGATION
Acting to reduce the scale of the next disaster: land-use zoning, hazard-resistant buildings and infrastructure

RECOVERY
Rebuilding infrastructure and services, rehabilitating injured (physically and mentally) peoples and their lives

Figure 10 The hazard management cycle

The 'Recovery' stage of the hazard management cycle might be thought of as the 'returning to normal' stage. This can happen after a few months but in some cases it takes years. The recovery stage depends on:

- the magnitude of the disaster — bigger means longer
- development level — lower means longer, as poorer people are more severely affected
- governance, because well-governed places will divert resources more effectively to recovery efforts
- external help, i.e. aid and financing to help the recovery effort.

The importance of recovery can be seen on Park's Model: the disaster response curve. This well-known model provides a simple visual illustration of the impact of a disaster. It can be used to compare different hazard events as shown by curves A–C on Figure 11.

- Curve A shows a disaster with a relatively small impact on quality of life and a short response phase. Quality of life begins to improve quickly and returns to normal within a few months.
- Curve B shows quality of life is impacted more than in A and reconstruction takes longer, but mitigation improves quality of life, meaning the community is better prepared for the next hazard.

Exam tip

Be prepared to compare and contrast examples in the exam, using concepts such as Park's Model as a framework for your answer.

■ Curve C shows a disaster with a major impact on quality of life, and a slow reconstruction phase. Even years later quality of life has not returned to levels before the disaster.

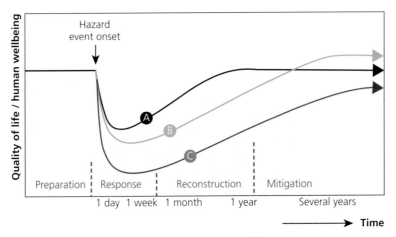

Figure 11 Park's Model
Source: after Park (1991)

It could be argued that the 2010 earthquake in Haiti fits profile C because the devastation was still not 'fixed' by 2015 and incomes, health, housing and food supply remained worse than pre-disaster. Developed countries are more likely to correspond to profiles A or B, developing ones to C.

Disaster modification

Disasters can be managed by modifying impacts. This can be done in three ways (Figure 12) by modifying the event, modifying vulnerability or modifying loss.

Event modification is the most desirable type of management but it is not always possible. Loss modification implies that a disaster has occurred and caused damage to people and property: this is the least desirable form of management.

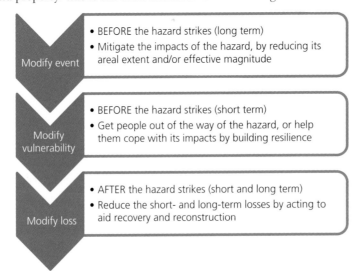

Figure 12 Disaster modification

Knowledge check 14

Which stage of the hazard management cycle would involve rebuilding schools, hospitals and businesses?

Event modification (Table 14) relies on technology and planning systems and can be high cost. It is less likely to be used in developing and emerging countries, although low-cost examples of **hazard resistant design** are possible to build.

Table 14 Modify the event

Type of modification and example	Advantages	Disadvantages
Land-use zoning: preventing building on low-lying coasts (tsunami), close to volcanoes and areas of high ground-shaking and liquefaction risk	▪ Low cost ▪ Removes people from high-risk areas	▪ Prevents economic development on some high-value land, e.g. coastal tourism ▪ Requires strict, and enforced, planning rules
Aseismic buildings: cross-bracing, counter-weights and deep foundations prevent earthquake damage	▪ Widely used technology can prevent collapse ▪ Protects both people and property	▪ High costs for tall/large structures ▪ Older buildings and low-income homes are rarely protected (Figure 13)
Tsunami defences: tsunami sea walls and breakwaters prevent waves travelling inland	▪ Dramatically reduces damage ▪ Provides a sense of security	▪ Can be overtopped ▪ Very high cost ▪ Ugly and restrict use/development at the coast
Lava diversion: channels, barriers and water cooling used to divert and/or slow lava	▪ Diverts the lava out of harms way ▪ Relatively low cost	▪ Only works for low VEI basaltic lava ▪ The majority of 'killer' volcanoes are not of this type

'Quake-proof' house

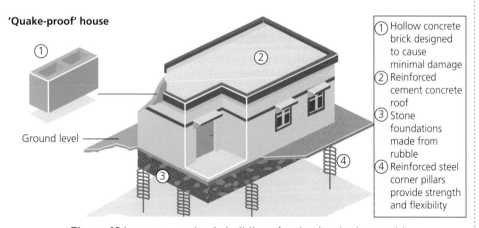

(1) Hollow concrete brick designed to cause minimal damage
(2) Reinforced cement concrete roof
(3) Stone foundations made from rubble
(4) Reinforced steel corner pillars provide strength and flexibility

Figure 13 Low-cost aseismic buildings for the developing world

Modifying vulnerability (Table 15) means increasing the resilience of a community to increase their capacity to cope. In many cases, prediction, warning and evacuation are used to move people out of harm's way. Predictions need to be accurate or there is a risk that '**cry wolf syndrome**' will reduce the effectiveness of warnings.

Hazard resistant design involves constructing buildings and infrastructure that are strong enough to resist tectonic hazards. In the case of earthquakes these are called aseismic buildings.

Cry wolf syndrome occurs when predictions (and evacuation) prove to be wrong, so that people are less likely to believe the next prediction and warning, and therefore fail to evacuate.

Table 15 Modify the vulnerability

Type of modification and example	Advantages	Disadvantages
Hi-tech scientific monitoring: used to monitor volcano behaviour and predict eruptions	■ In most cases, predicting an eruption is possible ■ Warnings and evacuation save lives	■ Costly, so not all developing world volcanoes are monitored ■ May suffer from 'cry wolf syndrome' ■ Does not prevent property damage
Community preparedness and education: 'earthquake kits' and preparation days, education in schools	■ Low cost, often implemented by NGOs ■ Can save lives through small actions	■ Does not prevent property damage ■ Harder to implement in isolated rural areas
Adaptation: moving out of harm's way and relocating to a safe area	■ Would save both lives and property	■ High population densities prevent it ■ Disrupts people's traditional homes and traditions

Loss modification could be described as 'picking up the pieces' after a disaster has occurred (Table 16). If event and vulnerability modification has also been used then the losses should be quite small. However, in the case of developing countries, loss modification is often the main management strategy. This was the case after the 2010 Haiti earthquake and 2004 Indian Ocean tsunami. In these cases management should be considered as having failed to protect people.

Table 16 Modify the loss

Type of modification and example	Advantages	Disadvantages
Short-term emergency aid: search and rescue followed by emergency food, water and shelter	■ Reduces death toll by saving lives and keeping people alive until longer-term help arrives	■ High costs and technical difficulties in isolated areas ■ Emergency services are limited and poorly equipped in developing countries
Long-term aid: reconstruction plans to rebuild an area and possibly improve resilience	■ Reconstruction can 'build in' resilience through land-use planning and better construction methods	■ Very high costs ■ Needs are quickly forgotten by the media after the initial disaster
Insurance: compensation given to people to replace their losses	■ Allows people to recover economically, by paying for reconstruction	■ Does not save lives ■ Few people in the developing world have insurance

Knowledge check 15

What type of disaster modification is least likely to save lives?

....

Earthquake kits are boxes of essential household supplies (water, food, battery powered radio, blankets) kept in a safe place at home to be used in the days following an earthquake.

Exam tip

Remember that different types of modification are more applicable to some tectonic hazards than others, and cannot always be used in the developing world because of cost restrictions.

....

Summary

- The distribution of earthquakes, volcanoes and tsunami is related to plate margins (divergent, convergent and conservative) and explained by the theory of plate tectonics.
- Tectonic processes result in a number of hazard types, including crustal fracturing, ground shaking, liquefaction and landslides, as well as a variety of volcanic hazards and tsunami.
- Natural hazards and disasters are not the same: disasters happen when a vulnerable population with low resilience experiences the negative impacts of natural hazards.
- Social, economic and environmental impacts vary by level of development as well as type of hazard process.
- Magnitude, frequency and the hazard profile of events are all important in terms of understanding variation in impacts.
- The urban, economic, demographic, political governance and social context of an area are all important in understanding vulnerability and community resilience.
- Trends show that while deaths and numbers affected by disasters are falling, economic damage is rising.
- Mega-disasters and multiple hazard zones are particularly important in terms of large-scale, sometimes global, impacts.
- There is variation in the extent to which tectonic hazards can be predicted.
- Models such as Parks Model and the Hazard Management Cycle can be useful in terms of understanding short- and long-term impacts, response, recovery and mitigation.
- Tectonic hazards can be managed by modifying the event, vulnerability or loss — each approach has advantages and disadvantages.

■ Landscape systems, processes and change

Glaciated landscapes and change

How has climate change influenced the formation of glaciated landscapes over time?

- The Earth's climate has fluctuated between warm interglacial and cold glacial conditions many times in the last few millions of years, on different timescales and with different causes.
- Ice cover today is very different to ice cover that was present in the past.
- Periglacial areas, at the edges of areas of ice cover on land, have also seen their distribution change over time.

Causes of climate change

Earth's **climate** has changed significantly in the past. Throughout Earth's history there have been two dominant climates:

1 icehouse conditions, when ice cover has been present across large areas
2 greenhouse conditions when the climate has been much warmer and largely ice free.

For the last 30 million years, icehouse conditions have prevailed. The last 2.6 million years are referred to as the Quaternary geological period. This is subdivided into two epochs:

1 the Pleistocene, from 2.6 million to 12,000 years ago
2 the Holocene, from 12,000 years ago to the present day.

The Pleistocene saw more than 20 major climate fluctuations where Earth's climate flipped between:

- interglacial periods, with warm temperatures similar to Earth's climate today
- glacial periods, with much colder temperatures compared with today and extensive ice cover.

Each cold–warm cycle lasted around 100,000 years.

Natural causes

The accepted main cause of climate change in the Pleistocene is changes in the Earth's orbit around the sun (orbital changes or astronomical forcing). This is based on Milankovitch cycle theory developed by Milutin Milankovitch in 1924. Milankovitch argued that the surface temperature of the Earth changes over time because the Earth's orbit and axis vary over time. These variations lead to changes in the amount and distribution of solar radiation received by Earth from the sun.

Climate refers to the 30-year average temperature and precipitation conditions for an area. It is distinct from weather, which refers to day-to-day changes in conditions.

Exam tip

Learn the approximate dates of geological time periods such as the Holocene and Pleistocene.

- On a timescale of 100,000 years, Earth's orbit changes from circular, to **elliptical**, and back again (called orbital eccentricity). This changes the amount of radiation received from the sun.
- On a timescale of 41,000 years Earth's axis tilts from 21.5° to 24.5° and back again (axial tilt). This changes the seasonality of Earth's climate. The smaller the tilt, the smaller the difference between summer and winter.
- Lastly, on a 22,000 year timescale, Earth's axis 'wobbles' and this changes the point in the year that the Earth is closest to the sun (axial precession).

Figure 14 shows the combined effects of these three cycles. Milankovitch's theory is supported by the fact that glacial periods have occurred at regular 100,000 year intervals.

An **elliptical** orbit is one shaped like a rugby ball, rather than a football (a circular orbit).

Knowledge check 16

Which of the three orbital changes has the largest impact on climate change?
....

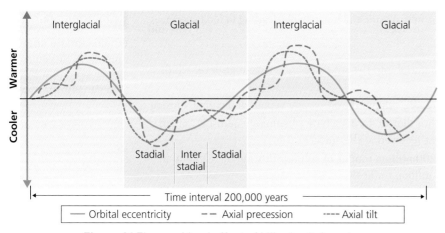

Figure 14 The combined effect of Milankovitch cycles

The impact of orbital changes on solar radiation amount and distribution is small. In total, it can probably change global temperature by ±0.5°C. The evidence of past climate change from ice cores and other evidence suggests that glacial periods were 5–7°C colder than interglacials. Scientists think that orbital changes may have been just enough to trigger a major global climate change, but climate **feedback mechanisms** are needed to sustain it.

- An example of positive feedback is snow and ice cover. Small increases in snow and ice dramatically raise surface **albedo**, and reflect solar energy back into space. This contributes to further cooling, which might encourage further snowfall, and therefore a cycle of cooling. This may be how a 0.5°C cooling caused by orbital changes could be amplified into a 5°C global cooling.
- An example of negative feedback is cloud cover. As warming occurs, more evaporation will occur and this may increase global cloud cover. Increasingly cloudier skies could reflect more solar energy back into space, and diminish the effect of the warming.

Solar output

The amount of energy emitted by the sun varies as a result of **sunspots**. The effect of sunspots is to blast more solar radiation towards the Earth. There is a well-known 11-year sunspot cycle, as well as longer cycles. The total variation in solar radiation caused by sunspots is about 0.1%. Sunspots have been recorded for around 2000 years, and there is a good record for around 400 years.

Feedback mechanisms are those than can either amplify a small change and make it larger (positive feedback), or diminish the change and make it smaller (negative feedback).

Albedo means the reflectivity of a surface. Snow and ice have a high albedo, reflecting back heat energy, whereas oceans and forests have a low albedo and absorb heat energy.

Sunspots are dark spots that appear on the sun's surface, caused by intense magnetic storms.

A long period with almost no sunspots, known as the Maunder Minimum, occurred between 1645 and 1715, and this is often linked to the Little Ice Age (Figure 15). Similarly, the Medieval Warm Period has been linked to more intense sunspot activity, although it is unclear whether the Medieval Warm Period was a global event. The Little Ice Age climate cooling event:

- lasted from 1450 to 1850
- experienced especially cold periods within that time period, at around 1660, 1770 and 1850
- had average temperatures that were 0.5–1°C colder than the 20th century
- caused rivers and lakes to freeze more regularly and sea ice in the Arctic to be more extensive
- made the growing season for farmers shorter and less reliable, leading to food shortages.

Volcanic eruptions

Major volcanic eruptions eject material into the stratosphere where high-level winds distribute it around the globe.

- Volcanoes eject huge volumes of ash, sulphur dioxide, water vapour and carbon dioxide.
- High in the atmosphere, sulphur dioxide forms a haze of sulphate aerosols, which reduces the amount of sunlight received at the surface.
- Tambora, Indonesia, ejected 200 million tonnes of sulphur dioxide in 1815 and in 1991 Mt Pinatubo ejected 17 million tonnes.

Tambora led to the 'year without a summer' in 1816 as global temperatures dipped by 0.4–0.7°C. Temperature dropped by 0.6°C following the Pinatubo eruption. These are short-lived effects, as the sulphate aerosols only persist for 2–3 years.

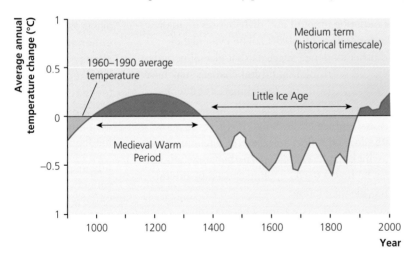

Figure 15 Climate change in the UK since AD 900

Loch Lomond Stadial

During the Pleistocene, at the end of the last glacial period (called the Devensian) when ice sheets were melting, there was a short but severe return to very cold conditions in the North Atlantic. This Loch Lomond Stadial event began 12,700 years ago and 'ice age' conditions returned. Glaciers began to grow in the Scottish

Highlands. Temperatures in the British Isles ranged from 10°C in summer to −20°C in winter. After about 1300 years, temperatures suddenly rose and have been warm ever since (see Figure 16). The most probable explanation is a sudden influx of cold freshwater to the North Atlantic caused by the melting ice sheets. This disrupted the **Thermohaline Circulation** in the North Atlantic and this sudden influx ended only when the supply of glacial meltwater ran out. This event tells us that natural climate changes can occur very rapidly.

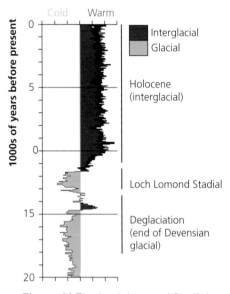

Figure 16 The Loch Lomond Stadial

Past and present ice cover

Ice cover found on planet Earth is referred to as the **cryosphere**. Ice is found in a wide range of different environments that broadly divide into:

- high latitude: ice found with the Arctic and Antarctic Circles, more than 65 degrees of latitude north and south
- high altitude: ice found in mountain ranges, which can occur at any latitude.

There are a number of different types of ice mass, shown in Table 17 on page 32.

Ice masses can be constrained or unconstrained. Ice sheets and ice caps blanket the topography beneath them (unconstrained) whereas ice fields and glacier edges butt-up against valley walls and mountain sides (constrained). Ice is also found in the sea, as ice shelves. This floating sea ice generates icebergs when wave action carves off chunks of the ice.

Ice masses can also be classified by **thermal regime**.

- Warm-based (found in temperate locations) glaciers are not frozen to the surface they sit on, i.e. the bedrock. At their base there is a layer of water, allowing the base of the ice to slide.
- Cold-based glaciers (found in polar high latitudes) have very cold surface temperatures of −20–30°C and their base is frozen to the bedrock beneath. Neither the pressure of the ice above nor geothermal energy from below is enough to melt the ice base.
- Polythermal glaciers have a cold-based upper portion at high altitude but a warm-based lower portion at lower altitude.

The **Thermohaline Circulation** is a system of interconnected ocean currents that help redistribute heat around the Earth, which helps even-out temperature differences between the equator and the poles.

Exam tip

It can be useful to sketch graphs, such as Figures 15 and 16, in the exam as it may save time.

Knowledge check 18

What was the last glacial period called?

The **cryosphere** is the discontinuous layer of ice covering parts of the Earth's land and ocean surface, as well as ice found underground.

Thermal regime refers to the temperature at the base of an ice mass, and determines whether there is ice or water present.

Knowledge check 19

What is the thermal regime of glaciers that slide on a layer of water at their base?

Content Guidance

Table 17 Types of ice mass

Ice mass type	Description	Area (km²)	Examples
Ice sheet	Large, dome-shaped ice masses that completely submerge underlying topography and are many kilometres thick	More than 50,000	Antarctica and Greenland
Ice cap	A smaller version of an ice sheet, with its high point corresponding to the high point of submerged mountains beneath the ice	Less than 50,000	Austfonna on Svalbard
Ice field	Ice covering a mountain plateau, but not thick enough to bury all of the topography or extend beyond the high altitude area	Less than 50,000	Garden of Eden ice field, New Zealand
Valley glacier	A long tongue of ice confined between valley sides and terminating in the valley or at the sea. Outlet glaciers drain ice caps and ice sheets	5–1500	Arolla Glacier (Switzerland/ Italy)
Cirque glacier	Smaller glacier occupying a hollow in a mountain side and circular in shape (also called a corrie glacier)	0.5–10	Hodges Glacier Gritvken, S Georgia

Distribution of ice cover

The cryosphere plays a key role in a number of global systems.

- It is a key component of the hydrological cycle, with about 69% of the world's fresh water locked up in ice. Ice captures water from the atmosphere and glacier melt releases this fresh water into river systems.
- Surface ice has a significant cooling effect on the planet, because of ice's high albedo. This reflects solar radiation back into space, cooling polar regions and regulating Earth's climate.
- Polar regions, especially **permafrost** areas, are significant global carbon stores, as biological carbon is locked away in frozen ground, preventing bacterial decay. This helps regulate the level of carbon dioxide in the atmosphere.

Today ice cover can be found on land and in the ocean, as shown in Table 18.

Table 18 The extent of ice on land and in the oceans

Ice on land	Percentage of global land surface
Antarctic ice sheet	8.3
Greenland ice sheet	1.2
Glaciers, ice caps and ice fields	0.5
Permafrost	9–12
Seasonal permafrost	33
Ice in the ocean	**Percentage of global sea surface**
Antarctic ice shelves	0.5
Sea ice (winter maximum)	9.1
Sub-sea permafrost	0.8

Exam tip

Learn definitions of each of the types of ice mass, as well as named examples of them.

Permafrost is soil, sediment or rock that has been frozen for two or more years.

32　Edexcel Geography

The vast majority of ice is in either permafrost or found in Antarctica. Glaciers, ice caps and ice fields are a very small proportion of the total ice cover. Today there are around 170,000 valley glaciers and numerous ice caps in the world but most are found in a small number of locations.

At the height of the last glacial period in the Pleistocene, ice covered about 32% of the total land area, compared with about 10% today (Figure 17). Ice sheets covered a huge area of North America, Northern Europe and north and central Asia as well as the Andes. In addition:

- sea ice extent was much larger, extending as far south as Japan and as far north as Argentina
- sea level was much lower, because so much water was locked up in land-based ice.

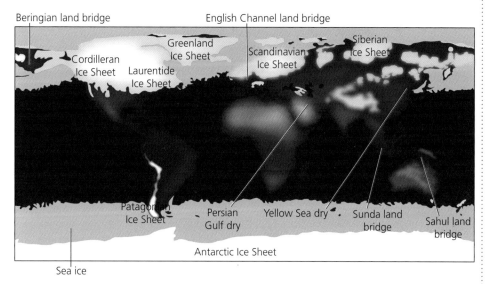

Figure 17 Maximum ice extent during the last glacial period, 30,000 years ago

There is extensive evidence of past high-altitude ice cover in the form of **relict landscapes** from the Pleistocene. The UK has many upland areas that were formerly glaciated. These include the Highlands and Grampians, the Lake District and Snowdonia.

Periglacial environments

Locations at the edge of ice-covered areas are called periglacial environments. In these places the defining feature of the landscape is that the ground is frozen — so-called permafrost. Periglacial environments are found:

- at high latitudes, especially in the Arctic. They are less common in the Southern Hemisphere because of the absence of land masses between 60°S and 70°S
- at high altitude because temperatures fall by 1°C for every 100 m of height, e.g. the Himalaya Plateau
- in continental interiors, away from the moderating influence of the sea, e.g. Siberia.

Relict landscapes are wholly or partly preserved landscapes, which show physical landform evidence for very different climate, environmental and physical process conditions in the past.

Figure 18 shows ice cover and permafrost extent in the Northern Hemisphere. It should be noted that permafrost extends out under the sea bed in the Arctic Ocean.

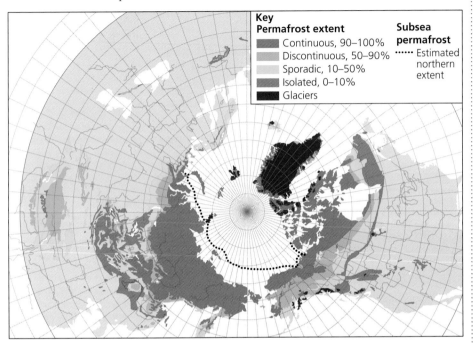

Figure 18 The extent of permafrost in the Northern Hemisphere

Source: Permafrost Laboratory, Geophysical Institute, University of Alaska, 2007

There are different types of permafrost:

- continuous permafrost (90%+ of the ground surface is underlain by permafrost) in areas where mean annual soil temperature is below −5°C
- discontinuous permafrost (50–90% of the ground surface) found in areas where the mean annual temperature is below −1°C
- sporadic permafrost (less than 10–50% of the ground surface) exists as patches, often in areas shaded from direct sunlight or where soils are more easily frozen, and is termed 'isolated' when the percentage cover falls below 10%.

Freezing extends downwards to the point where **geothermal heat** from below prevents freezing. This can be deeper than 1000 m. Permafrost has a summer seasonally active layer. This is the upper layer which seasonally melts. The depth varies from a few centimetres to several metres, depending on the average summer temperature.

In the past, the area of permafrost was much more extensive. At the height of the last glacial period, permafrost covered most of southern England and the landscape was one of **tundra**.

Periglacial processes

Periglacial areas are affected by a number of physical processes, many of which are highly seasonal and dependent on annual cycles of freezing and thawing. These processes contribute to a distinctive periglacial landscape and landforms in places such as northern Russia and northern Canada (Table 19).

Geothermal heat is heat conducted from within the Earth towards the surface.

The **tundra** is the ecosystem or biome found in most periglacial areas (alpine tundra if at high altitude). It has no trees, low rainfall and consists mostly of low-growing grassland plants adapted to very cold and dry conditions.

Table 19 Periglacial processes and landforms

Process	Explanation	Related landforms
Nivation	Seasonal snow collects in shallow depressions and moves slowly downslope	Nivation hollows: weathering, erosion and meltwater flow caused by the snow's movement deepen the depression into a bowl-shaped hollow
Frost heave	The upward swelling of soil due to the growth of ice lenses within the soil. The ice lenses are fed by capillary action or groundwater flow from below. Frost heave also forms patterned ground	Pingos: these are circular, soil covered mounds up to 70 m high which have an ice core, usually fed by a groundwater source from below allowing the ice core to swell upward forcing the soil mound upward
Thermal contraction	The shrinkage of the ground surface due to extreme low temperatures, creating cracks in the surface with a regular, interlocking pattern of polygons (patterned ground)	Ice wedge polygons: ice wedges from when spring meltwater pools in thermal contraction cracks, only to refreeze when it hits permafrost below. Over many seasons the ice wedge grows in thickness
Freeze–thaw weathering	On freezing, water expands by 9% in volume. Repeated cycles of freezing and thawing of water in cracks in rock splits rocks into progressively smaller blocks	Blockfields and scree: frost-shattering, angular rock fragments can form on a flat surface as a blockfield, or at the base of a slope, forming scree
Solifluction	A type of mass movement, or soil creep, found on low-angle slopes: saturated soil of the active layer flows very slowly downslope in summer but freezes in the winter	Solifluction terraces and lobes: solifluction produces vegetated lobes or terraces giving a slope a 'stepped' profile
Wind erosion and deposition	Periglacial areas often have strong winds, and these can erode the finely ground rock debris produced by glaciers and ice caps	Loess fields: extensive areas of wind-blown glacial sediment found at the margins of ice

Exam tip

You need to be able to recognise the landforms in Table 19 in photographs and diagrams, as well as explain their formation.

Patterned ground is a unique feature of periglacial areas consisting of polygon, net and stripe patterns on the ground surface, made either of stones or ice wedges.

Knowledge check 21

Which process, in periglacial areas, breaks rocks down into smaller angular fragments?

A final feature of periglacial areas is the extreme seasonality of their river regimes. Rivers have no discharge in winter because of freezing conditions, but this can change rapidly in summer as seasonal melting begins, leading to extensive (but short-lived) meltwater erosion on river banks.

What processes operate within glacier systems?

- Glaciers operate as systems, and the concept of mass balance is important in understanding the system.
- Ice masses move in different ways and at different rates.
- The action of ice creates unique landscapes as a result of erosion, entrainment, transport and deposition processes.

Glaciers as systems

Glacial ice begins to form when snowfall from one winter lasts until the next winter. Further snowfall begins the process of building multiple layers of ice. The basic process is one of compression as the weight of new snow above compresses old snow below.

A **river regime** is the change in discharge experienced by a river over the course of a year, its seasonal flow variation.

- **Firn** forms after about 2 years, as snow recrystallises into sugar-grain sized crystals with small pore spaces between them.
- Ice forms slowly, over 20–30 years in some cases, as further compression and recrystallisation takes place.

In very cold places such as Antarctica the transformation of snow to ice may take hundreds of years.

Ice masses are a system of inputs (snow), transfers (ice movement) and outputs (meltwater). The snow and **meltwater** can be gained (accumulation) and lost (ablation) in a number of ways (Table 20).

Accumulation	Ice movement	Ablation
• Direct snowfall onto the ice mass • Avalanches carrying snow onto an ice mass from upslope • Windblown snow being moved from another area and deposited onto an ice mass	→	• Melting of ice at the margins of the ice mass • Calving of ice from the margin on an ice mass where it meets the sea • Evaporation and sublimation from the ice mass surface • Avalanches from the ice mass itself

Table 20 Ice masses as systems

For an ice mass to form accumulation has to be greater than ablation. This occurs:
- in a high altitude environment, such as a mountain valley or depression in the landscape
- in a high latitude location, where average annual temperatures are very low.

As a mass of ice accumulates it will begin to flow away from the zone of accumulation. Eventually the ice mass will reach an area where temperatures are higher (a lower altitude and/or latitude) and ablation will exceed accumulation.

Glacier mass balance

The balance between ice mass inputs and outputs is called mass balance. This is shown diagrammatically in Figure 19. In a valley glacier such as the one in Figure 19, the mass balance is in equilibrium if accumulation and ablation cancel each other out over the course of a year. Figure 20 shows that over the course of a year there is a period of net accumulation (winter) and net ablation (summer) but over the whole year the system is in balance.

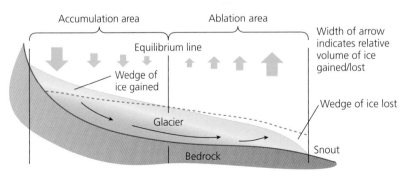

Figure 19 Glacier mass balance

Firn is the initial stage of ice formation, with a density of 550–830 kg/m³ (compared with 500 kg/m³ for snow or névé. Ice has a density of about 900 kg/m³).

Exam tip

Many glacial processes need to be explained as a step-by-step, logical sequence, using precise terminology.

Knowledge check 22

What is 'firn'?

Meltwater is the flow of water from streams and rivers at the snout of a glacier or edge of a larger ice mass.

Knowledge check 23

What term is used to describe the build up of snow on an ice mass, usually in the winter?

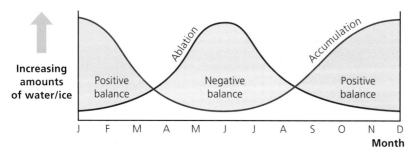

Figure 20 The annual ice budget of a glacier in the Northern Hemisphere

Exam tip

It is useful to be able to draw a quick sketch of a mass balance diagram, such as Figure 19, for exam questions on this topic.

Equilibrium means being in balance, i.e. equal amounts of accumulation and ablation over a year or other timescale.

Knowledge check 24

What is thought to be the main cause of the increased negative mass balance of glaciers seen worldwide in the last few decades?

Ice-streams are zones of fast-moving ice at the edges of ice sheets, which funnel ice towards the edges of the ice mass where it melts.

Exam tip

Ice movement is complex; make sure you learn the different processes of movement and the types of ice mass they affect.

Wasting refers to ice masses that are shrinking in size as a result of net ablation.

Lithology refers to different rock types and their characteristics.

Ice masses with net ablation will shrink over time, whereas ones with net accumulation will grow. Feedback mechanisms can have an impact on mass balance.

- A small increase in snowfall caused by a lowering of temperature can increase the albedo of the ice surface, contributing to further cooling and more accumulation (positive feedback).
- It is possible that higher temperatures in some areas can increase evaporation and therefore precipitation (snowfall) so more snow accumulation balances the losses from ablation in higher temperatures (negative feedback).

Figure 20 shows how mass balance changes on the timescale of a year. Mass balance also changes on longer timescales as ice masses respond to climate trends. Many Alpine glaciers grew during the Little Ice Age and their snouts extended lower down alpine valleys. Mass balance was often close to **equilibrium** (= 0) in the 1980s, but there has been an accelerating loss of ice since. This change is usually ascribed to the impact of global warming.

Glacier movement

Ice moves because of gravity. Although solid, ice can deform before it fractures so is often described as 'flowing', rather like a fluid. The velocity of moving ice varies from about 0.5 m per year at the centre of cold-based ice sheets to up to 10 m per *day* at the snouts of some glaciers.

In general, cold-based polar ice masses move slowly and warm-based temperate high-altitude glaciers flow faster. However, some **ice-streams** that drain polar ice sheets and caps move much faster than their parent ice mass.

There are several different mechanisms of ice movement, outlined in Table 21 (page 38). Ice velocity is dependent on temperature, as higher temperatures encourage basal sliding. Steep valley glaciers will have higher flow rates than ice sheets on relatively flat topography, and the high altitude parts of valley glaciers may be cold-based for all or part of the year, slowing velocity. In addition a glacier experiencing net ablation (negative mass balance) may be rapidly **wasting** and therefore flowing quickly. Ice velocity also changes along the long profile of a glacier as a result of bed topography (Figure 21). It may also change because of variations in **lithology**, with easily eroded rocks (sandstone, heavily fractured metamorphic rock) being more prone to bed deformation.

Table 21 How glacial ice moves

Basal slip	Ice slides across bedrock/sediment substrate because of a layer of meltwater between the ice and substrate. Water-saturated sediment beneath ice can also slide, carrying the ice with it (bed deformation)	Warm-based, temperate glaciers and the margins of some cold-based ice masses
Regelation creep	Occurs when ice melts under pressure, usually because it encounters an obstacle. The meltwater flows around the obstacle and refreezes as the pressure drops	Valley glaciers on steep, rocky slopes at the ice/substrate contact
Internal deformation	Movement between or within individual grains of ice, including grains slipping over each other, melting and recrystallising, and slip within individual grains (intra-granular slip)	The upper parts of ice masses. The only type of motion on cold-based ice masses

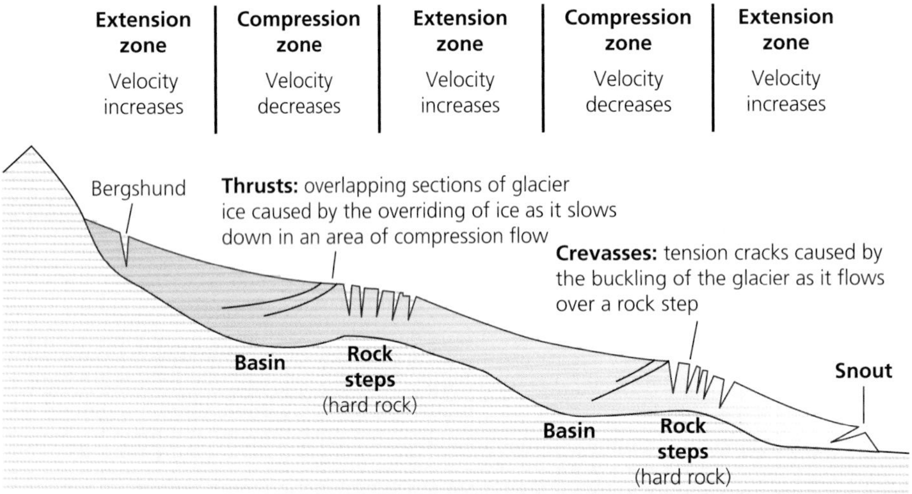

Extension zone	Compression zone	Extension zone	Compression zone	Extension zone
Velocity increases	Velocity decreases	Velocity increases	Velocity decreases	Velocity increases

Bergshund

Thrusts: overlapping sections of glacier ice caused by the overriding of ice as it slows down in an area of compression flow

Crevasses: tension cracks caused by the buckling of the glacier as it flows over a rock step

Basin

Rock steps (hard rock)

Basin

Rock steps (hard rock)

Snout

Figure 21 Extensional and compressional ice flow as a result of bed topography

Knowledge check 25

What is the main cause of ice movement in warm-based glaciers?

Glacier landform systems

It might seem strange to be studying ice masses when there are none in the UK today. But in geological terms the UK landscape was transformed by ice very recently — the ice only disappeared about 12,500 years ago. Figure 22 shows the variety of glacial and periglacial environments that once existed in Great Britain. These were formed by a combination of four processes.

1   Glacial erosion: the removal of rock material by ice through abrasion and plucking and by meltwater flow from ice margins.
2   Glacial **debris** entrainment: sediment being incorporated into glacial ice and carried along with the ice.
3   Glacial sediment transport: sediment on top of (supraglacial), within (englacial) or at the base (subglacial) of ice being moved.

Debris refers to any glacially eroded and transported sediment, of any size.

4 Glacial deposition: when material is released from the ice at the margin or the base of a glacier and simply dumped, or dumped out of meltwater.

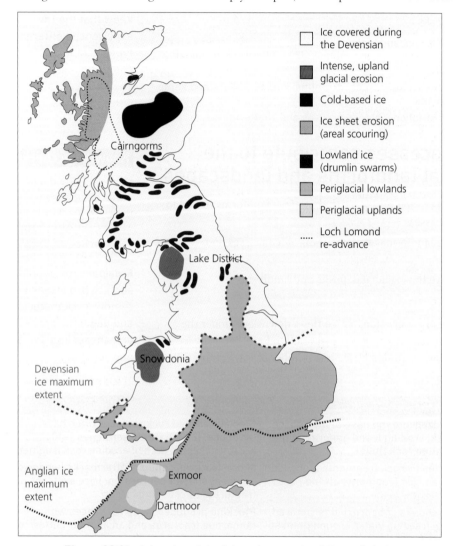

Legend:
- Ice covered during the Devensian
- Intense, upland glacial erosion
- Cold-based ice
- Ice sheet erosion (areal scouring)
- Lowland ice (drumlin swarms)
- Periglacial lowlands
- Periglacial uplands
- Loch Lomond re-advance

Figure 22 Glacial and periglacial environments of Great Britain

These processes occurred in four types of locations, evidence for all of which can be found in the UK landscape.

1 Subglacial areas, which were covered by either warm- or cold-based ice for tens of thousands of years at a time, are dominated by the action of erosion and the erosional landforms it produced.

2 Ice-marginal areas, at the edges of ice sheets and valley glaciers, are areas where deposition from ice was dominant.

3 Proglacial areas, in front of ice masses where meltwater deposition and the action of wind on glacially eroded sediments can be found.

4 Periglacial areas, which had no ice cover but were underlain by permafrost covered in tundra vegetation.

In these locations landforms, which vary in scale, provide evidence of former glacial and periglacial conditions, as shown in Table 22.

Table 22 Glacial landforms by scale

Macro-scale	Ice-sheet-eroded knock and lochan landscapes, cirques, arêtes and pyramidal peaks. Glacial troughs, ribbon lakes, till plains, terminal moraines, sandurs
Meso-scale	Crag and tail, roches moutonnées, drumlins, kames, eskers and kame terraces, kettle holes
Micro-scale	Features such as striations, glacial grooves and chattermarks, erratics

How do glacial processes contribute to the formation of glacial landforms and landscapes?

- There are a number of glacial erosion processes that produce landforms linked to valley glaciers and large ice masses.
- Depositional processes create landforms at the margins of ice masses and underneath them.
- Meltwater produces distinctive fluvioglacial deposits and landforms.

Glacial erosion

Ice masses are very powerful agents of **erosion**. As ice flows downslope under the influence of gravity it erodes rock material that is transported to lower levels by ice and then deposited. A rock surface can be eroded by ice in a number of ways (Table 23).

Table 23 Types of glacial erosion

Type	Explanation	Evidence
Abrasion	Rock fragments frozen into the base of moving ice grind down bedrock, eroding it and generating fine eroded material called rock flour	Glacially eroded rock surfaces often have striations: long, parallel scratches 1–10 mm wide, tracing the path of eroding rock fragments
Crushing	The pressure exerted by rock fragments embedded at the base of ice can chip fragments off the bedrock below	Micro-features called chattermarks (wedge or crescent shaped depressions) are evidence of fracturing by crushing forces
Plucking (quarrying)	Plucking removes chunks of bedrock (in fractured and jointed rock) by freezing water around them so they are pulled away by the ice mass as it moves over the top	Plucking produces steep rock faces with numerous fractures and an angular pattern of 'missing' blocks
Basal melting	This process is similar to abrasion in rivers, but differs because water flowing at the base of ice can be under pressure, making abrasion a more powerful force	Deeply eroded channels and potholes can indicate areas where basal meltwater flow occurred in the past

Erosion generates sediment that is incorporated into ice. Two other processes can also be sources of glacial sediment.

1 Freeze–thaw weathering on the rock slopes above valley glaciers is a constant source of angular rock fragments that build up to form scree.
2 Mass movement — in the form of rock falls and landslides — is a source of rock material which falls onto the surface of ice and is incorporated into it.

Erosion is the action of surface processes that remove rock material and begin the process of transporting it to a new location.

Mass movement is the downslope movement of material under the influence of gravity.

Distinctive landforms of erosion

The variety of ice masses and differences in erosion types produce a range of erosional landforms that reflect their environment of origin (Figure 23). Constrained cirque and valley glaciers erode an impressive landscape of steep, rocky landforms that can be best seen in active areas such as the Alps and relict areas such as the Lake District. Table 24 summarises these landforms.

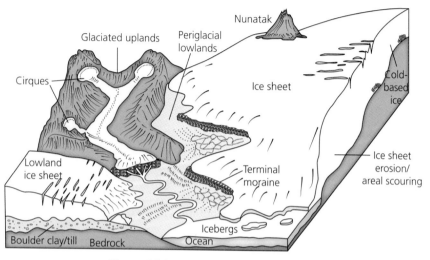

Figure 23 Different glacial environments

Exam tip

There are many different processes and landforms to learn for this section of the specification. Using flash cards will help you to revise them.

Knowledge check 27

Which erosion processes produce chattermarks and striations?

Exam tip

Make sure you are clear about the different landforms associated with valley glaciers and ice sheets.

Table 24 Cirque and valley glacier erosional landforms

Landform	Description	Formation processes
Cirque/corrie	Bowl-shaped depression with a steep, craggy backwall and rock or sediment lip, usually a few hundred metres in diameter	Rotation of an area of accumulated ice causing abrasion and plucking of the bedrock
Arête	A steep rock ridge separating two corries/cirques, formed by their backwalls	Lateral erosion of cirque backwalls gradually narrowing the rock between them
Pyramidal peak	A very steep rock peak in the shape of a pyramid	Formed when three or more cirque back walls and arêtes meet
Glacial trough/ribbon lake	The valley eroded by a glacier, being steep-sided but flat-bottomed (U-shaped). Many are several kilometres long and may today contain a long, thin lake (ribbon lake)	V-shaped river valleys are eroded by ice, and deepening occurs as most erosion is concentrated at the ice base
Truncated spurs	Spurs are projections of bedrock into a river valley, but they are cut (truncated) by glacial erosion to form steep rock faces	Ice shapes former river valleys by erosion cutting into the interlocking spurs of rock and slicing off their ends
Hanging valley	A small glacial valley cutting into a larger one from the side, with its base high above the floor of the larger valley	Ice from a smaller, less deep glacier joins a larger glacier so their bases are at different elevations

Ice sheet scouring

Ice sheets are unconstrained ice masses (Figure 23). At their centre they are often cold-based so very little erosion occurs, but towards their edges they are warm-based. Their huge extent and mass means that they 'sandpaper' vast areas by a process called areal scouring, leading to a relatively low topography landscape.

The west of Scotland, the northwest Highlands and the Isle of Lewis have a very irregular lowland landscape of small, bare-rock, ice-sculpted hills (knocks — similar to roche moutonnées) and small lakes (lochans). This is called 'knock and lochan' topography. This landscape is a result of:

- resistant bedrock over an extensive area, metamorphic gneiss in the west of Scotland
- widespread abrasion and plucking by an ice sheet moving in one direction for a long period of time, called areal scouring
- selective deeper erosion of faults, master joints and weaker areas of rock to form lochans
- areas of more resistant rock that form the higher knocks.

Knock and lochan landscapes often have a distinct regional alignment of small hills and lakes that follow lines of weakness, e.g. master joints or weaker rock strata. **Differential geology** can contribute to the formation of two erosional landforms, as shown in Figure 24.

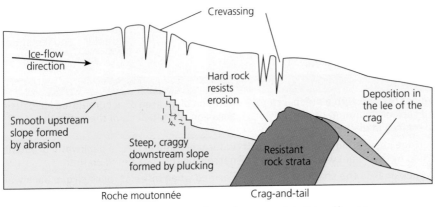

Figure 24 Roche moutonnée and crag-and-tail landforms

Glacial depositional landforms

The most common glacial depositional landforms are moraines. Moraine is **unconsolidated** glacial sediment and is unsorted (many different particle sizes) and angular in shape. Moraines are ice-contact landforms, meaning that the sediment has been transported only by ice since it was weathered and/or eroded. Four types of moraine are common.

1 Terminal moraines: these mark the maximum position of the ice, and are the heap of debris dumped by the snout of a glacier. In the case of valley glaciers they are arc-shaped. Ice sheet terminal moraines can be tens of metres high and many kilometres long.

2 Recessional moraines: similar to terminal moraines, but marking the positions of an ice mass as it gradually melted and retreated during deglaciation.

3 Lateral moraines: rock debris builds up along the edge of valley glaciers where the ice meets the valley wall, because freeze–thaw weathering and mass movement constantly drop debris onto the ice. When the ice melts the debris is dumped in a hummocky line along the valley edge.

4 Medial moraines: when two valley glaciers meet to form one larger glacier, their respective lateral moraines merge into a medial moraine running down the centre of the larger glacier and when this ice melts, this debris is deposited as a line of moraine running down the centre of the glacial trough.

Lowland features

Till is the unconsolidated sediment deposited by a glacier. It is usually a mix of clay, boulders and gravel:

- lodgement till is deposited under the ice and is usually structureless
- ablation till is deposited by melting ice and usually has some evidence of deposition in running meltwater.

Under lowland ice sheets (see Figure 23) there is often a layer of glacial sediment that forms a till plain when the ice melts. The till plain consists of a thick deposit often called boulder clay and it blankets the bedrock topography below. Much of northern England is covered in this type of deposit.

Till often forms a landscape of drumlins. A drumlin is a small, rounded hill — shaped like half an egg ('basket of eggs topography') (Figure 25). Drumlins often occur together in 'swarms' or 'fields'.

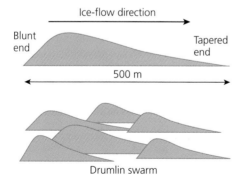

Figure 25 The characteristic shape of drumlins

> ### Knowledge check 29
>
> Look at Figure 24. What is the main difference between roche moutonnées and crag-and-tail landforms?

Drumlins:

- are lined up parallel to the direction of ice flow
- have their 'blunt' end facing into the ice-flow direction, and their tapered end facing downstream of the direction of ice flow
- are typically 100–2000 m long and 50–600 m wide
- are usually under 50 m high
- are made of glacially deposited sediments, although some have a rock core.

Drumlins are thought to have formed by deposition under moving ice sheets. As such, they are a subglacial feature. Most are made of lodgement till but some have fluvio-glacial deposits, so water might play a role in their formation. A process involving deformation (moulding and 'streamlining') of lodgement till by relatively fast-moving ice sheets is one way they may form. Moulding by meltwater between ice and till could be responsible for some of the fluvioglacial sediments found in some drumlins.

Ice extent and movement

Glacial landforms can be used to reconstruct the extent of ice cover in the past, and the direction ice masses moved in. This is very important for geographers when trying to reconstruct past glacial periods.

One problem is that the landforms created during one glacial period tend to be at best modified, and in many cases destroyed, by later glaciations. Most evidence in the UK is for the Devensian glaciation (110,000 to 12,000 years ago). Evidence for the earlier Anglian glaciation (400,000 to 320,000 years ago) is much rarer. Table 25 summarises the evidence used in ice-mass reconstruction.

Table 25 Evidence of former ice mass extent and movement

Ice extent	Ice movement
Terminal and recessional moraines Mark the forward edge of an ice mass so can be used to reconstruct its outer limit, but are vulnerable to subsequent erosion	**Erratics** Very useful for proving that ice flowed from one location to another if the erratic can be linked to its origin
Till plain/boulder clay cover May indicate ice extent, but only for warm-based ice. These deposits can be difficult to date accurately	**Drumlin orientation** The long axes of drumlins are aligned in the direction of ice flow, so can be used to show patterns of ice movement
Fluvioglacial deposits Most fluvioglacial action happened at the edge of ice masses so these can indicate areas that were not ice-covered or were at the ice margin	**Crag-and-tail and roche moutonnée** Both landforms have an 'upstream' end, so are good indicators of ice flow direction

Meltwater and fluvioglacial landforms

Many glacial depositional landforms are actually the result of the work of ice and water. This is not surprising as all ice eventually melts at the edge of an ice mass and flows away as meltwater. Within ice, meltwater flows in one of three locations, shown on Figure 26. Meltwater not only transports water through the ice mass but also carries sediment with it. This sediment is subject to the attrition processes sediment experiences in rivers so will become more rounded and smaller during its transport by meltwater.

Exam tip

There are many terms for glacial deposits, such as boulder clay, lodgement and ablation till; make sure you are using the correct one.

Fluvioglacial sediments are those that originate through ice erosion but are then transported, eroded and deposited by the meltwater from ice masses.

Erratics are rocks and boulders that were glacially eroded in one place, but transported by ice and deposited somewhere else. Their rock type does not match the area in which they are deposited.

Knowledge check 30

What is an 'erratic'?

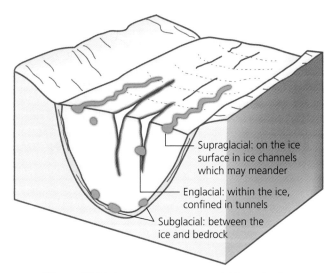

Figure 26 Meltwater movement within a glacier

Supraglacial: on the ice surface in ice channels which may meander

Englacial: within the ice, confined in tunnels

Subglacial: between the ice and bedrock

Glacial and fluvioglacial sediment deposits have very different characteristics because they formed with or without the action of water. These characteristics are summarised in Table 26. Figure 27 illustrates glacial and fluvioglacial sediments to highlight their differences.

Table 26 Glacial and fluvioglacial sediment characteristics

Characteristic	Explanation	Glacial	Fluvioglacial
Stratification	Stratified sediment has distinct layers within it. Unlayered sediment is described as 'massive', meaning no layers are present	Unstratified, massive	Stratified, representing different phases of deposition
Sorting	Well-sorted sediment has one dominant sediment size, e.g. most is sand-sized particles 1–2 mm in size. Poorly sorted sediments have a wide mix of sediment sizes	Poorly sorted	Well sorted
Imbrication	Imbricated sediments have their long axes (a-axis) aligned in one direction, usually indicating deposition from flowing water	Not imbricated: random axis orientation	Imbricated: dominant axis orientation
Shape	Shape ranges from angular to rounded: water rounds sediment during attrition whereas angular clasts have not been subject to the action of water	Angular	Rounded or subrounded, depending of length of time in transport
Grading	In running water, large sediment is deposited first as the water slows, and fine sediment last. This creates graded sediments with pebbles, sand, silt and clay being deposited in a sequence from bottom to top	Ungraded	Graded (if a range of clast sizes is present)

All sediment has three **axes**, a, b and c. These correspond to the length (a), width (b) and height (c) of a sediment particle.

Exam tip

You need to be able to recognise sediment characteristics from graphs, rose diagrams and in cross-section.

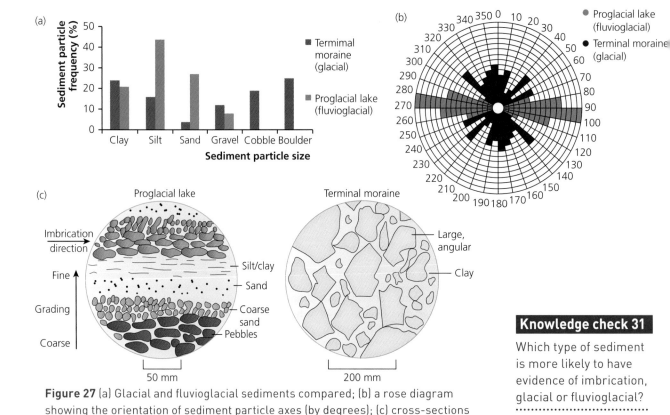

Figure 27 (a) Glacial and fluvioglacial sediments compared; (b) a rose diagram showing the orientation of sediment particle axes (by degrees); (c) cross-sections through proglacial lake and terminal moraine sediments

Knowledge check 31

Which type of sediment is more likely to have evidence of imbrication, glacial or fluvioglacial?

Fluvioglacial landforms

Water is responsible for landform formation in two situations:

1 englacial and subglacial meltwater flow helps produce landforms, which are termed ice-contact landforms

2 meltwater deposits sediment beyond the ice mass edge, as proglacial landforms.

Proglacial landforms are associated with glacial outwash plains.

Table 27 summarises these fluvioglacial landforms.

An **outwash plain** is the area in front of an ice mass where the dominant processes are those of meltwater erosion and deposition. Sandur is another name for a glacial outwash plain.

Table 27 Fluvioglacial landforms

	Landform	Description	Explanation
Ice contact	Kames and kettleholes	Mounds and hillocks of till, with some evidence of meltwater flow; often found next to depressions called kettle holes which may form small ponds and pools	Ice surface sediment is dumped as ice melts beneath it, and some reworking by meltwater occurs. Kettle holes are the depressions left as individual ice blocks melt
	Eskers	Long, thin, sinous gravel ridges (like embankments) often extending hundreds of metres	Formed in ice-walled englacial tunnels (deposited as the ice melts) or constrained subglacial streams
	Kame terrace	Sediment 'benches' on valley sides representing the former ice surface and its contact with the valley wall	Sediment seems to accumulate in ice-bound pools and exhibits stratification and is often well-sorted

	Landform	Description	Explanation
Proglacial	Sandhur	Extensive, flat sediment plain with numerous meltwater channels draining an ice mass	The dominant processes are river ones, i.e. channel formation and deposition of sediment from river flow, although they are seasonal because of summer ice melt
	Proglacial lake	A lake formed in a depression in front of an ice mass. They are often dammed by moraines	Sediment settles in seasonal layers called varves on the lake bed, and exhibits grading
	Meltwater channels	Braided channels and meandering channels running across the surface of a sandur	Formed by meltwater from the ice mass, but more mobile than most river channels because of the lack of stabilising vegetation

How are glaciated landscapes used and managed today?

- Glacial landscapes have high environmental, cultural, economic and scientific value to a range of people.
- There are local and global threats which are damaging fragile active and relict glacial landscapes.
- Managing glaciated landscapes to reduce the threats they face is possible but requires action at local, national and international scales.

The value of glaciated landscapes

Many glacial landscapes are relatively isolated areas with low population densities. This includes:

- active high altitude landscapes such as the Himalaya and Alps
- active high latitude landscapes including the periglacial tundra of northern Russia and Canada
- relict upland glacial landscapes such as the Lake District and Cairngorms in the UK.

These areas share some common features, including:

- highly seasonal climates, even in the relict landscapes
- geographical isolation and difficult access
- traditional peoples, often still farming and using resources in a traditional way.

Glaciated landscapes can have strong religious and spiritual associations, giving them a special significance and cultural value to indigenous people (Table 28).

Table 28 Spiritual and religious associations

Himalayan landscape	Arctic Inuit religion
■ Mount Kailash, which has never been climbed, is sacred to Buddhism, Jainism and Hinduism ■ Tibetan prayer flags are often placed on mountains, passes and ridges — showing the importance of the natural world in Buddhism ■ In Tibet, pilgrimages take place every year to sacred mountains and lakes	■ All things, living and non-living have a spirit, a form of animism ■ Inuit gods and goddesses often control a particular environmental realm ■ Examples include Sedna (goddess of sea animals) and Nanook (god of polar bears) ■ Many Inuit religious stories and myths warn against physical dangers in the Arctic

Braided channels are a network of numerous small and large channels that interweave, rather than there being one dominant channel. They are common in proglacial areas because of large variations in meltwater flow and lack of vegetation stabilising channels' banks.

Indigenous people are the original inhabitants of a territory and have strong historical and cultural connections with that area.

Knowledge check 32

What types of landscape features often have religious significance in glaciated landscapes?

Glaciated landscapes are important scientifically for a number of reasons.

- Polar regions are ideal locations for studying the upper atmosphere, as the atmosphere is thinner at the poles and less polluted than elsewhere on Earth.
- The dark skies and clear air of polar and high-altitude regions make astronomical research easier than in other places.
- Scientific research has indicated that the Arctic and Antarctic act as 'barometers' for global warming. Their pristine condition and low human population mean that environmental change shows up in these regions first and can be studied.

However, because of globalisation and the increasing availability of transport to once-isolated glacial regions, they are becoming more and more popular with visitors. Arctic and Antarctic cruises, polar trekking, and extreme wilderness sports and recreation are increasingly common in glacial areas that were inaccessible only 30 years ago.

Exam tip

Make sure you know why glaciated landscapes are valuable culturally, scientifically and economically.

Economic activity and importance

Despite low populations a wide range of economic activity takes place in glaciated landscapes. However, the type of landscape has a major impact on economic activity, as can be seen by a comparison of the economic activities found in the Lake District and Greenland.

Lake District: relict upland glacial

- Farming, especially upland hill sheep farming, which generates low incomes for farmers.
- Tourism is a major income earner, especially in summer months, reflecting the beauty of the landscape and its National Park status.
- Forestry is important in some locations, because of the Forestry Commission's commercial soft-wood plantations such as Whinlatter.

Greenland: active ice sheet margin

- The largest industry is fishing for cod, prawns and halibut — farming is very limited because of the extreme cold.
- Tourism is of growing importance and includes whale watching and 'iceberg' tourism but it is highly seasonal.
- Mining and oil exploration are growing in importance as technology allows exploitation in harsh environments and global warming makes Greenland more accessible.
- Greenland has built hydroelectric power (HEP) plants in Nuuk, Sisimiut and Ilulissat, providing the towns with electricity.

Biodiversity and natural systems

In many glaciated landscapes, both active and relict, **biodiversity** is relatively low by global standards. The periglacial Arctic tundra vegetation consists of dwarf shrubs, mosses, lichens, grasses and sedges. As with plant and animal species everywhere, tundra species may contain important genetic material and/or chemical compounds that could benefit food, pharmaceutical or engineering science.

Arctic ice cover and periglacial areas, and the Antarctic, actually play crucial roles in global physical systems despite their apparent isolation. Table 29 explores this.

Biodiversity is the number of plant and animal species in a given area.

Knowledge check 33

Which groups of plant and animal species does the Arctic tundra have a high percentage of their world total?

Table 29 The Arctic and global natural systems

The biosphere: animal migration	The climate system: global refrigeration
Migration is an important component of Arctic ecology. Many bird species including ducks and geese fly north in the spring to breed in the short Arctic summer, and then migrate south in the winter. Caribou undertake similar, but shorter, migrations. Overall species health depends on these migrations	The heat deficit of polar regions is balanced globally by atmospheric circulation and the thermohaline circulation that moves excess tropical heat towards the poles. The high albedo of Arctic sea ice, snow cover and glacier ice reflects 85–90% of incoming solar energy back into space and this has a significant cooling effect on the planet
The water cycle: ice stores	**The carbon cycle: sequestration**
Water is stored in glacial ice and is gradually released as meltwater, both into the oceans and rivers. This is an important component of the hydrological cycle. In some locations, such as the Himalaya, summer meltwater from glaciers is a critical component in human water supply	Wetlands, peat lands and lakes cover up to 70% of the Arctic. These ecosystems are an important carbon sink as they store un-decomposed organic matter as well as protecting permafrost. Arctic soils, permafrost and seabeds also store methane. Some estimates suggest up to 1400 gigatonnes of methane is stored in the Arctic

Threats to glacial landscapes

Glacial environments are locations of natural hazards but, given their low populations, the impacts of natural hazards tend to be small. There are exceptions. One of these is avalanches. Figure 28 shows numbers of deaths from avalanches in the USA since 1950. There is a clear rising trend, which can be attributed to:

- increasing numbers of skiers in high altitude, glacial areas
- the rising popularity of off-piste and extreme skiing
- increased accessibility and improved equipment, encouraging people to venture into isolated areas.

An **avalanche** is a mass movement (landslide) involving snow, ice and rock. Most occur on slopes of 35–45° and the majority are triggered by changes in weather or people on the slope.

Exam tip

Make sure you can link high latitude glaciated landscapes to wider global systems, including the hydrosphere, atmosphere and biosphere.

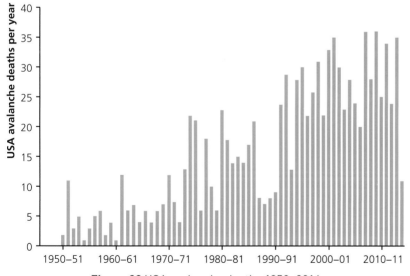

Figure 28 USA avalanche deaths 1950–2014

Glacial outburst floods, or jökulhlaups in Icelandic, are also a threat. These are uncontrolled releases of large volumes of meltwater from underneath ice masses. Often they are triggered by geothermal heating or volcanic eruptions beneath ice sheets and glaciers. While rare, jökulhlaups can cause major damage through flooding and erosion.

Exam tip

Take care with your spelling, especially when words are of Icelandic origin!

- The 1996 eruption in Iceland of Grímsvötn volcano under the Vatnajökull glacier caused a jökulhlaup with a flood flow rate of 50,000 cubic metres per second, which destroyed parts of Iceland's main ring road.
- The flood carried ice blocks weighing over 100 tonnes and the initial flood front was 4 m high.
- In total about 180 million tonnes of sediment were moved by the flood waters.

These floods are fairly common in the Himalaya region, affecting Bhutan, Tibet, Nepal and Pakistan — often as a result of ice-dammed lakes suddenly bursting.

Upland glacial landscapes are subject to a range of human pressures and threats that can degrade their landscape and fragile ecology (Figure 29). Tourism can have a major impact:

- 120 million tourists visit the Alps each year
- over 300 ski resorts have been developed in the Alps
- extreme sports such as mountain-biking, canyoning and paragliding are encroaching on areas previously untouched by tourism
- air pollution levels are rising as vehicle numbers in the Alps increase year on year.

Knowledge check 34

What is a jökulhlaup?

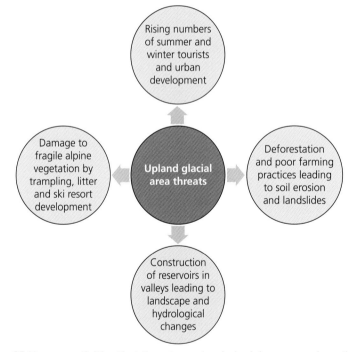

Figure 29 Human activities that threaten upland glacial areas such as the Alps

The narrow but deep glacial troughs in upland glacial landscapes are ideal for dam and reservoir construction. HEP stations in dams can provide power to urban areas and regulate water supply. But dam and reservoir building dramatically alters the landscape and **hydrologic regime** of rivers. The vast Himalaya area is especially vulnerable to this threat, as Table 30 shows.

The **hydrologic regime** of a river refers to its seasonal variations in discharge, both of water and sediment.

Table 30 Constructed and planned HEP dams in the Himalaya

	Nepal	Pakistan	Bhutan	Indian Himalaya region
Built	15	6	5	74
Under construction	2	7	—	37
Planned	37	35	16	318

Global warming

Rising global temperatures represent the biggest threat to ice masses because:

- higher temperatures lead to negative mass balances and shrinking glaciers, ice caps and ice sheets
- periglacial areas will experience melting permafrost and dramatic changes to the ecosystem and biodiversity of the tundra.

In human terms, the biggest consequence of these changes is likely to be an alteration to the hydrological cycle. In many parts of the world people depend on glacial meltwater for their water supply. In the Peruvian Andes some 30% of water comes from this source. As glaciers shrink, so may water supply. However, this may be many decades away because increased melting could temporarily increase supply. There are other possible consequences:

- increased flood risk if summer ice melt spikes because of extreme heat conditions and causes increases in **river discharge**
- low meltwater levels limiting HEP dam operation
- low meltwater levels failing to 'flush' rivers clean, leading to higher concentrations of pollution and lower water quality
- changes in the **sediment yield** of glacial meltwater streams — either becoming choked with excess sediment because of low flow (and increasing the risk of flooding) or dramatically rising during melting and outburst events risking lives and property.

Managing threats to glacial landscapes

Threats to glacial landscapes can be managed in a number of ways, depending on the aims and attitudes of those players responsible — the stakeholders (Table 31).

Table 31 The spectrum of approaches to protection

Protection	Sustainable management	Multiple economic use
Wildlife reserves and wilderness areas designed to limit human activity and maximise landscape and ecological protection	National parks usually try to balance conservation and the needs (economic, social and cultural) of people living in the area	Management may prioritise economic development, especially if it is limited and incomes are low. Conservation is a secondary consideration
The Arctic National Wildlife Refuge (ANWR) in the USA is an example: human activity is limited and oil exploitation is banned. Antarctica, under the Antarctic Treaty, is the ultimate example of this approach. The ANWR is the most biodiverse region in the Arctic	UK national parks such as the Lake District use this method. Economic development is allowed but in a managed, low-key way and the use of the 'Sandford Principle' ensures that conservation has priority over development if the two aims conflict	Greenland's development of mineral resources, fishing, tourism and HEP (partly using global warming as an economic opportunity) may risk damage to its fragile ice-sheet, tundra and ocean environments in a region where economic development opportunities are traditionally very limited

River discharge is the volume of water flowing down a river, usually measured in cubic metres per second.

Sediment yield is the volume and type of sediment carried by a river at any one time.

Knowledge check 35

In what way is glacial meltwater important in terms of human quality of life?

Exam tip

You need to know examples of different types of protection in glaciated landscapes, both active and relict.

The twin aims of conservation and landscape protection versus economic development often conflict, and decisions need to be made in terms of which gets priority. Figure 30 summarises the roles of different stakeholders involved in managing glaciated landscapes.

Figure 30 Stakeholders involved in managing glaciated landscapes

Legislative frameworks

A key challenge of managing glacial landscapes is when the landscape requiring protection spans across international borders. This is the case in the Alps, Andes, Himalaya, Arctic and Antarctic. The challenges of this situation are:

- countries may have contrasting attitudes towards conservation, management and exploitation of resources
- legal systems may be different between countries
- financial and human resources available for management may be very different
- countries may have poor international relations, making agreement hard to reach.

Two examples of legislative frameworks are summarised in Table 32. The Antarctic Treaty is a special case. This is because there is no indigenous population in the Antarctic and territory is not owned by any nation (territorial claims have been set aside as part of the Antarctic Treaty).

A **legislative framework** is a set of laws, agreements and treaties (if international) that govern actions by different stakeholders in order to protect and conserve an area.

Table 32 Examples of legislative frameworks

Alpine Convention	Antarctic Treaty
■ Intergovernmental treaty that came into force in 1995 ■ It includes the EU plus Austria, Germany, France, Italy, Liechtenstein, Monaco, Slovenia and Switzerland ■ Attempts to balance development in the Alps with environmental protection ■ Encourages member states to develop policies to manage planning, air pollution, soil conservation, water management, conservation, farming, forests and tourism	■ Covers the entire area south of 60° latitude ■ Came into force in 1961 and has 52 member states ■ Bans military activity in Antarctica ■ Guarantees freedom to use Antarctica for scientific research ■ It also limits tourism to day visits only, and prevents permanent settlement (other than scientific bases) ■ Fishing is strictly managed, and mineral exploitation is banned in Antarctica

Co-ordinated approaches

The most significant threat to glacial environments is climate change. The threat can be summarised as:

■ a high risk of 6–8°C warming in the Arctic by 2100, meaning large areas of periglacial permafrost would melt, releasing millions of tonnes of stored carbon

■ 95% of the world's glaciers have a negative mass balance and are retreating, and many will disappear by 2100

■ accelerating melting of the Greenland ice sheet

■ uncertainty about how the Antarctic ice sheet will respond to global warming.

International legislative frameworks and local protection such as national parks can only go so far in protecting glacial environments from the **context risk** of global warming. These actions need to be co-ordinated with a global agreement to reduce greenhouse gas emissions to slow, or stop, global warming. The 1987 Montreal Protocol and 1997 Kyoto Protocol are both examples of such actions, but they only involved developed countries. In December 2015 agreement was reached at the COP21 Paris UN climate change conference to reduce emissions. Figure 31 shows what was agreed in Paris. If pledged cuts were implemented, emissions would slow, but not enough to limit global warming to the +2°C many scientists argue is required to avoid major changes to glacial environments. Successful management of these unique and fragile landscapes is increasingly challenging, with a need for co-ordinated approaches at global, national and local scales.

Further co-ordinated action is likely to be needed to protect the fragile cryosphere.

Exam tip

Be aware that the Antarctic Treaty is a special case — no other area on Earth is managed in quite the same way.

A **context risk** is one that affects the whole world, such as global warming, so any management actions need to be taken in the context of this wider risk.

Knowledge check 36

Explain why global warming is a context risk.

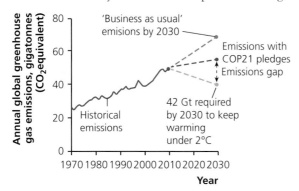

Figure 31 Global greenhouse gas emissions (carbon dioxide equivalent)

Content Guidance

Summary

- There have been long-term changes to climate and ice cover (glacial and interglacial periods) and shorter-term changes during the Quaternary and Holocene.
- Milankovitch cycles, solar variation and volcanic eruptions are all potential causes of changes to climate and ice cover.
- The cryosphere can be found in high latitude and high altitude locations today, but its distribution was different in the past.
- Periglacial environments are widespread in high latitudes today, but also had a different distribution in the past.
- Glacier mass balance is an important concept to understand the balance of accumulation and ablation of an ice mass.
- Ice masses move by different processes and at different rates, which influences the landscapes ice produces.
- Both glacial erosion, deposition and fluvioglacial processes generate specific landforms that together produce contrasting active and relict glacial landscapes.
- Relict and active glacial landscapes are important culturally, economically, and in terms of biodiversity and the water and carbon cycles.
- Human activity, including global warming and more local threats, risks degrading glacial landscapes, which can also experience natural hazards.
- A wide variety of different players are involved in managing glaciated landscapes, using a spectrum of approaches with different levels of protection both locally and using international legislative frameworks.

■ Landscape systems, processes and change

Coastal landscapes and change

Why are coastal landscapes different and what processes cause these differences?

- The coastal zone, or littoral zone, has a wide variety of distinctive features and landscapes.
- Geological structure is important in generating different coastal landscapes and features.
- Geology, and other factors, is important in affecting rates of coastal retreat.

The coastal zone

Over 1 billion people live on coasts at risk from flooding and about 50% of the world's population live within 200 km of the coast. Many live in the littoral zone, which in many cases is a dynamic area of high risk. Risks include coastal flooding and coastal erosion. The littoral zone (Figure 32) can be divided into:

- coast: land adjacent to the sea and often heavily populated and urbanised
- backshore zone: above high tide level and only affected by waves during exceptionally high tides and major storms
- foreshore: where wave processes occur between the high and low tide marks
- nearshore: shallow water areas close to land and used extensively for fishing, coastal trade and leisure
- offshore: the open sea.

The littoral zone is the wider coastal zone, including coastal land areas and shallow parts of the sea just offshore.

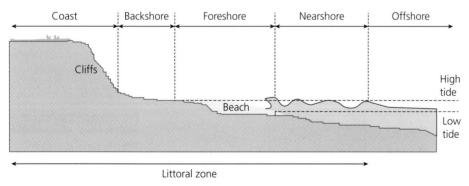

Figure 32 The littoral zone and its subzones

Classifying coastal landscapes

Coasts can be divided into two main types:

1 rocky (or cliffed) coastlines with cliffs varying in height from a few metres to hundreds of metres

2 coastal plains (with no cliffs) where the land gently slopes towards the sea across an area of deposited sediment, often in the form of sand dunes and mud flats.

Cliffs create a sharp distinction between 'land' and 'sea' whereas coastal plains gradually transition from land to sea.

Coasts can be classified according to different physical features and processes (Table 33).

Table 33 Different coastal classifications

Formation processes	Relative sea level change	Tidal range	Wave energy
Primary coasts are dominated by land-based processes, such as deposition at the coast from rivers or new coastal land formed from lava flows	**Emergent coasts** where the coast is rising relative to sea level, for example as a result of tectonic uplift	Tidal range varies hugely on coastlines, meaning coasts can be: ■ Microtidal (tidal range of 0–2 m) ■ Mesotidal (tidal range of 2–4 m) ■ Macrotidal (tidal range greater than 4 m)	**Low energy** sheltered coasts with limited fetch and low wind speeds resulting in small waves
Secondary coasts are dominated by marine erosion or deposition processes	**Submergent coasts** are being flooded by the sea either because of sea level rise and/or subsiding land		**High energy** exposed coasts, facing prevailing winds with long wave fetches resulting in powerful waves

Rocky coasts

Many coastlines consist of rocky cliffs that vary in height from a few metres (low relief) to hundreds of metres (high relief). High relief cliffs are composed of relatively hard rock. There are two main cliff profile types (Figure 33).

1 Marine erosion dominated: wave action dominates and cliffs tend to be steep, unvegetated and there is little rock debris at the base of the cliff.

2 Subaerial process dominated: not actively eroded at the base by waves; shallower, curved slope and lower relief; surface runoff erosion and mass movement are responsible for the cliff shape.

Marine erosion dominated

Subaerial process dominated

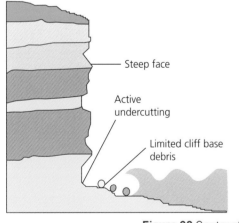
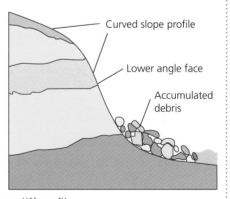

Steep face

Active undercutting

Limited cliff base debris

Curved slope profile

Lower angle face

Accumulated debris

Figure 33 Contrasting cliff profiles

Coastal plains

Low-lying, flat (low-relief) areas close to the coast are called coastal plains. Many contain estuary wetlands and marshes, being just above sea level and poorly drained because of the flatness of the landscape. Coastal plains form when:

- sea level falls, exposing the sea bed of what was once a shallow continental shelf sea, e.g. the Atlantic coastal plain in the USA
- sediment brought from the land by river systems is deposited at the coast causing coastal **accretion** so coastlines gradually move seaward, such as a river delta
- sediment is moved from offshore sources (sand bars) towards the coast by ocean currents.

Coastal plains are a low-energy environment usually lacking large and powerful waves except on rare occasions such as during hurricanes.

> **Accretion** refers to the deposition of sediment at a coast that expands the area of land.

Geological structure

Geological structure is the arrangement of rocks in three dimensions. There are three elements to geological structure.

1 Strata: the different layers of rock exposed in a cliff.
2 Deformation: tilting and folding by tectonic activity.
3 Faulting: major fractures that have moved rocks from their original positions.

Geological structure produces two main types of coast:

1 concordant, or Pacific coasts, when rock strata run parallel to the coastline
2 discordant, or Atlantic coasts, when different rock strata intersect the coast at an angle, so rock type varies along the coastline.

Discordant coastlines are dominated by headlands and bays. Less resistant rocks are eroded to form bays whereas more resistant geology remains as headlands protruding into the sea. Figure 34 shows the West Cork coast in Ireland.

> **Exam tip**
>
> You will need an example of both a discordant and a concordant coast for the exam.

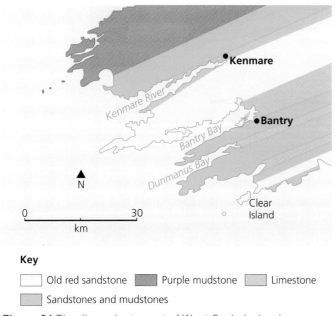

Key

☐ Old red sandstone ▨ Purple mudstone ▨ Limestone

▨ Sandstones and mudstones

Figure 34 The discordant coast of West Cork, Ireland

In this area:

- rock strata meet the coast at 90° in parallel bands
- weak rocks have been eroded, creating elongated, narrow bays
- more resistant rocks form headlands
- especially resistant areas remain as detached islands, such as Clear Island.

Headlands and bays change over time because headlands are eroded more than the bays. This happens because of the effect these coasts have on wave crests:

- in deep water wave crests are parallel
- as water shallows towards the coast waves slow down and wave height increases
- in bays, wave crests curve to fill the bay and wave height decreases
- the straight wave crests refract, becoming curved, spreading out in bays and concentrating on headlands
- the effect of **wave refraction** is to concentrate powerful waves at headlands (so greater erosion) and create lower, diverging wave crests in bays so reducing erosion.

Concordant coasts are more complex. Different rock strata run parallel to the coast but vary in terms of their resistance to the sea. The most well-known example is Lulworth Cove (Figure 35), where:

- the hard Portland limestone and fairly resistant Purbeck Beds protect much softer rocks landward (the Wealden and Gault beds)
- marine erosion has broken through the resistant beds, and then rapidly eroded a wide cove behind
- there is resistant chalk at the back of these coves, which prevents erosion further inland.

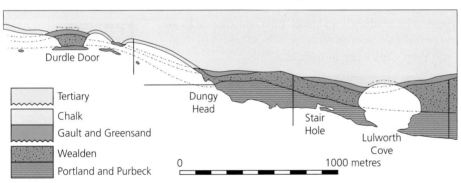

Figure 35 Lulworth Cove geology

The Dalmatian coast in the Adriatic Sea is another example of a concordant coastline, where:

- the geology is limestone
- the limestone has been folded by tectonic activity into a series of **anticlines** and **synclines**, that trend parallel to the modern coastline.

This underlying structure of upstanding anticlines and lower syncline basins (which would have been eroded by rivers in the past) has been drowned by sea level rise to create a concordant coastline of long, narrow islands arranged in lines offshore. Haff coastlines are also concordant. They are found on the southern edge of the Baltic Sea. Long sediment ridges topped by sand dunes run parallel to the coast just offshore, creating lagoons (the haffs) between the ridges and the shoreline.

Knowledge check 38

What is the technical name for the different layers of rock found in cliffs?
....

Wave refraction is the process causing wave crests to become curved as they approach a coastline.

Anticlines and **synclines** are types of geological fold caused by tectonic compression. Anticlines form crests and synclines form troughs.

Cliff profiles

Cliff profiles are influenced by geology. Two characteristics are dominant:

1 the resistance to erosion of the rock

2 the dip of rock strata in relation to the coastline.

Dip, meaning the angle of rock strata in relation to the horizontal, is important. Dip is a tectonic feature. Sedimentary rocks are formed in horizontal layers but can be tilted by tectonic forces. When this is exposed on a cliffed coastline it has a dramatic effect on cliff profiles, as shown in Figure 36.

> A **cliff profile** is the height and angle of a cliff face, plus its features such as wave-cut notches or changes in slope angle.

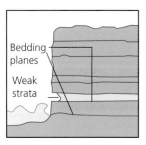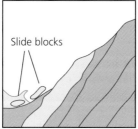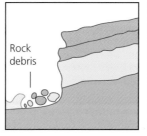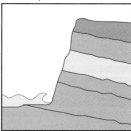

Horizontal dip	Seaward dip high angle	Seaward dip low angle	Landward dip
Vertical or near-vertical profile with notches reflecting strata that are more easily eroded	Sloping, low-angle profile with one rock layer facing the sea; vulnerable to rock slides down the dip slope	Profile may exceed 90° producing areas of overhanging rock; very vulnerable to rock falls	Steep profiles on 70–80° and producing a very stable cliff with reduced rock falls

Figure 36 Geology and cliff profiles

Other geological features influence cliff profiles and rates of erosion. These include:

- **faults**: either side of a fault line, rocks are often heavily fractured and broken and these weaknesses are exploited by marine erosion
- joints: occur in most rocks, often in regular patterns, dividing rock strata up into blocks with a regular shape
- fissures: much smaller cracks in rocks, often only a few centimetres or millimetres long but they also represent weaknesses that erosion can exploit
- folded rocks: often heavily fissured and jointed, meaning they are more easily eroded.

The location of **micro-features** found within cliffs, such as caves and wave-cut notches, are often controlled by the location of faults and/or strata which have a particularly high density of joints and fissures.

> **Faults** are major fractures in rocks produced by tectonic forces and involve displacement of rocks on either side of the fault line.
>
> **Micro-features** are small-scale coastal features such as caves and wave-cut notches which form part of a cliff profile.

Knowledge check 39

Name a type of coastal landform that often forms at the base of a cliff because of a weakness in the rocks, such as a fault line.

Coastal recession

Coastal recession refers to how fast a coastline is moving inland. This is influenced by many factors, but the key one is lithology or rock type. The three major rock types — igneous, sedimentary and metamorphic — erode at different rates, as shown in Table 34.

Table 34 Geology and coastal recession rates

Rock type	Examples	Erosion rate and explanation
Igneous	Granite Basalt Dolerite	**Very slow** Less than 0.1 cm per year Igneous rocks are crystalline, and the interlocking crystals make for strong, hard erosion-resistant rock Igneous rocks such as granite often have few joints, so there are limited weaknesses that erosion can exploit
Metamorphic	Slate Schist Marble	**Slow** 0.1–0.3 cm per year Crystalline metamorphic rocks are resistant to erosion Many metamorphic rocks exhibit a feature called foliation, where all the crystals are orientated in one direction, which produces weaknesses Metamorphic rocks are often folded and heavily fractured, forming weaknesses that erosion can exploit
Sedimentary	Sandstone Limestone Shale	**Moderate to fast** 0.5–10 cm per year Most sedimentary rocks are clastic and eroded faster than crystalline igneous and metamorphic rocks The age of sedimentary rocks is important, as geologically young rocks tend to be weaker Rocks with many bedding planes and fractures, such as shale, are often most vulnerable to erosion

Where the rock forming cliffs is **unconsolidated material** (such as sand or boulder clay) rates of recession can be much greater. Boulder clay on the Yorkshire Holderness coast erodes at 2–10 m per year.

Erosion and weathering resistance are influenced by:

- how reactive minerals in the rock are when exposed to chemical weathering
- whether rocks are **clastic** or **crystalline** — the latter are more erosion-resistant
- the degree to which rocks have cracks, fractures and fissures, which are weaknesses exploited by weathering and erosion.

Many cliffed coastlines are made of different rock types and so have complex cliff profiles (Figure 37) and experience differential erosion of alternating strata. Cliff profiles can also be influenced by the permeability of strata.

- Permeable rocks allow water to flow through them, and include many sandstones and limestones.
- Impermeable rocks do not allow groundwater flow and include clays, mudstones and most igneous and metamorphic rocks.

Permeability is important because groundwater flow through rock layers can weaken rocks by removing the cement that binds sediment in the rock together. It can also create high **pore water pressure** within cliffs, which affects their stability. Water emerging from below ground onto a cliff face at a spring can run down the cliff face and cause surface runoff erosion, weakening the cliff.

Exam tip

Become a bit of a geologist, so you can talk with confidence about different rock types and their properties.

Unconsolidated material is sediment that has not been cemented to form solid rock, a process known as lithification.

Clastic rocks consist of sediment particles cemented together, whereas **crystalline** rocks are made up of interlocking mineral crystals.

Pore water pressure is an internal force within cliffs exerted by the mass of groundwater within permeable rocks.

Knowledge check 40

Is sandstone a clastic or crystalline rock?

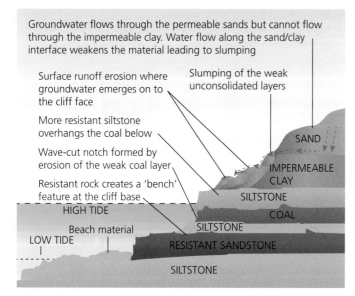

Groundwater flows through the permeable sands but cannot flow through the impermeable clay. Water flow along the sand/clay interface weakens the material leading to slumping

Surface runoff erosion where groundwater emerges on to the cliff face

Slumping of the weak unconsolidated layers

More resistant siltstone overhangs the coal below

Wave-cut notch formed by erosion of the weak coal layer

Resistant rock creates a 'bench' feature at the cliff base

HIGH TIDE

Beach material

LOW TIDE

SILTSTONE

RESISTANT SANDSTONE

SILTSTONE

SILTSTONE

COAL

IMPERMEABLE CLAY

SAND

Figure 37 A complex cliff profile

Coastal vegetation

Some coastlines including coastal sand dunes, salt marshes and mangrove swamps, are protected from erosion by the stabilising influence of plants. Vegetation stabilises coastal sediment in a number of ways.

- Roots bind sediment particles together making them harder to erode.
- When submerged, plants provide a protective layer so the sediment surface is not directly exposed to moving water and erosion.
- Plants protect sediment from erosion by wind, by reducing wind speed at the surface because of friction with the vegetation.

Many plants that grow in coastal environments are specially adapted halophytes (salt tolerant) or xerophytes (drought tolerant).

Plant succession

On a coast where there is a supply of sediment and deposition takes place:

- pioneer plant species will begin to grow in the bare sand or mud
- this forms the first stage of **plant succession**
- each step in plant succession is called a seral stage
- the end result of plant succession is called a climatic climax community.

Coastal climax communities include sand dune ecosystems (psammosere) and salt marsh ecosystems (halosere).

On coastal dunes the succession begins with the colonisation of embryo dunes by pioneer species (Figure 38). Embryo dune pioneer plants:

- stabilise the mobile sand by their root systems
- reduce wind speeds at the sand surface, allowing more sand to be deposited
- add dead organic matter to the sand, beginning the process of soil formation.

Knowledge check 41

How do plants contribute to preventing erosion of sediment on coasts?

Plant succession means the changing structure of a plant community over time as an area of initially bare sediment is colonised.

Embryo dunes alter the environmental conditions from harsh, salty, mobile sand to an environment that other plants can tolerate. New plant species therefore colonise the embryo dunes, creating a fore dune.

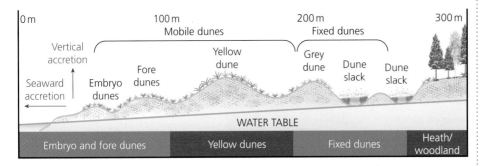

Sea	Bare sand	Sea rocket Saltwort Couch grass Lyme grass	Marram grass	Marram grass Sedge grass	Red fescue Heather Creeping willow	Bramble Pine Birch

Figure 38 Sand dune plant succession

A similar process of successional development happens on bare mud deposited in estuaries at the mouths of rivers that are exposed at low tide but submerged at high tide. Estuarine areas are ideal for the development of salt marshes because:

- they are sheltered from strong waves, so sediment (mud and silt) can be deposited
- rivers transport a supply of sediment to the river mouth, which may be added to by sediment flowing into the estuary at high tide.

How do characteristic coastal landforms contribute to coastal landscapes?

- Wave types and erosion processes are important in terms of the production of coastal landforms.
- Sediment transport and deposition processes produce coastal landforms, often stabilised by plant succession.
- The sediment cell concept shows how coasts operate as holistic systems.
- Weathering and mass movement subaerial processes are important on some coastlines.

Marine processes and waves

Waves are caused by friction between wind and water transferring energy from the wind into the water. The force of wind blowing on the surface of water generates ripples, which grow into waves when the wind is sustained. In open sea:

- waves are simply energy moving through water
- the water itself only moves up and down, not horizontally
- there is some orbital water particle motion within the wave, but no net forward water particle motion.

Wave size depends on a number of factors:

- the strength of the wind
- the duration the wind blows for
- water depth
- wave **fetch**.

Waves break as the water depth shallows towards a coastline (Figure 39).

- At a water depth of approximately half the wavelength, the internal orbital motion of water within the wave touches the sea bed.
- This creates friction between the wave and the sea bed, and this slows down the wave.
- As waves approach the shore, wavelength decreases and wave height increases, so waves 'bunch' together.
- The wave crest begins to move forward much faster than the wave trough.
- Eventually the wave crest outruns the trough and the wave topples forward (breaks).

Fetch is the uninterrupted distance across water over which a wind blows, and therefore the distance waves have to grow in size.

Exam tip

The process of waves breaking is a classic physical process sequence, which is best explained in logical steps using short sentences.

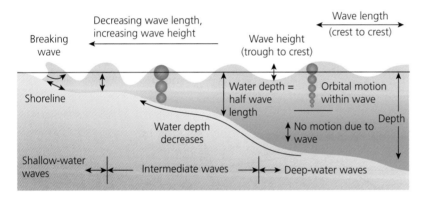

Figure 39 Breaking waves

Knowledge check 42

At what water depth do waves begin to break?

Constructive and destructive waves

Waves breaking on a shoreline do not all have the same shape. There is a basic difference between constructive and destructive waves (Figure 40).

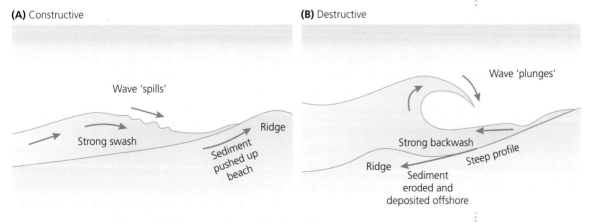

Figure 40 Constructive and destructive waves

Constructive (spilling or surging) waves:

- have a low wave height (less than 1 m) and long wavelength of up to 100 m
- are 'flat' waves with a strong swash but weak backwash
- have strong **swash** that pushes sediment up the beach, depositing it as a ridge of sediment (berm) at the top of the beach
- have a **backwash** that drains into the beach sediment.

Destructive (plunging) waves:

- have a wave height of over 1 m and a wavelength of around 20 m
- have strong backwash that erodes beach material and carries it offshore, creating an offshore ridge or berm.

Depending on conditions, beaches experience both constructive and destructive waves over the course of time and this can mean significant changes to **beach morphology** (beach sediment profile) on different timescales:

- over a day, as a storm passes and destructive waves change to constructive ones as the wind drops
- between summer and winter (Figure 41)
- when there are changes to climate, e.g. if global warming resulted in the UK climate becoming on average stormier, then destructive waves and 'winter' beach profiles would become more common.

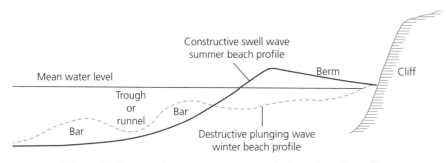

Figure 41 Changes between summer and winter beach profiles

Beaches have landforms that change constantly.

- Storm beaches, high at the back of the beach, result from high energy deposition of very coarse sediment during the most severe storms.
- Berm ridges, typically of shingle/gravel result from summer swell wave deposition.
- Low channels and runnels between berms.
- Offshore ridges/bars formed by destructive wave erosion and subsequent deposition of sand and shingle offshore.

Swash is the flow of water up a beach with a breaking wave.

Backwash is the water draining down the beach back into the sea. Both can transport sediment.

Beach morphology means the shape of a beach, including its width and slope (the beach profile) and features such as berms, ridges and runnels. It also includes the type of sediment (shingle, sand, mud) found at different locations on the beach.

Exam tip

Figure 39 is a good example of a diagram worth knowing for the exam. Explaining the beach profile differences in prose would be very time-consuming.

Marine erosion processes

Waves cause erosion, but erosion is not a constant process. Most erosion occurs during a small number of large storms. There are four main processes that cause erosion (Table 35).

Table 35 Erosion processes

Process name	Explanation
Hydraulic action (wave quarrying)	Air trapped in cracks and fissures is compressed by the force of waves crashing against the cliff face Pressure forces cracks open, meaning more air is trapped and greater force is experienced in the next cycle of compression This process dislodges blocks of rock from the cliff face
Abrasion (corrasion)	Sediment picked up by breaking waves is thrown against the cliff face The sediment acts on the cliff like a tool, chiselling away at the surface and gradually wearing it down
Attrition	As sediment is moved around by waves, the numerous collisions between particles slowly chip fragments off the sediment, making it smaller and more rounded over time
Corrosion (solution)	Carbonate rocks (limestones) are vulnerable to solution by rainwater, spray from the sea and seawater

Erosional coastal landforms

Erosion produces a suite of distinctive coastal landforms. The most well known is the cave–arch–stack–stump sequence (Figure 42). The most fundamental process of landform formation on a coastline is the creation of a wave-cut notch. This is eroded at the base of a cliff by hydraulic action and abrasion.

- As the notch becomes deeper, the overhanging rock above becomes unstable and eventually collapses as a rock fall.
- Repeated cycles of notch-cutting and collapse cause cliffs to recede inland.
- The former cliff position is shown by a horizontal rock platform visible at low tide, called a wave-cut platform.

Some landforms are influenced by structural geology. The location of fault lines and major fissures influences the location of caves and therefore arches.

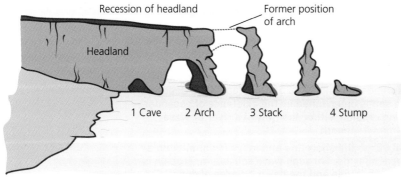

Figure 42 The cave–arch–stack–stump sequence

Sediment transport and deposition

Material eroded from cliffs is transported by the sea as sediment. Transport of sediment happens in four ways (Table 36).

Table 36 Sediment transport processes

Traction	Sediment rolls along, pushed by waves and currents	Pebbles, cobbles, boulders
Saltation	Sediment bounces along, either because of the force of water or of wind	Sand-sized particles
Suspension	Sediment is carried in the water column	Silt and clay particles
Solution	Dissolved material is carried in the water as a solution	Chemical compounds in solution

Waves breaking at 90° to a coast move sediment up and down beaches, but sediment can be transported along the coastline too. The process of sediment transport along the coast is called **longshore drift**. When wave crests approach a coast at an angle (rather than at 90° to the coast) the swash from the breaking waves, and the subsequent backwash, follow different angles up and down the beach in a zig-zag pattern. On most coastlines there is a dominant prevailing wind so over time there is a dominant direction of longshore drift.

Longshore drift is the net transport of sediment along the beach as a result of sediment transport in the swash and backwash.

Depositional landforms

Longshore drift is a key source of sediment for depositional landforms on coasts. Sediment transported down river systems to the coast or from offshore sources is also important. Sediment is deposited when the force transporting sediment drops. Deposition can occur in two main ways.

1 Gravity settling occurs when the energy of transporting water becomes too low to move sediment. Large sediment will be deposited first followed by smaller sediment (pebbles → sand → silt).

2 Flocculation is a depositional process that is important for very small particles, such as clay, which are so small they will remain suspended in water. Clay particles clump together through electrical or chemical attraction, and become large enough to sink.

Knowledge check 44

Name the process that transports sediment along the coastline and contributes to many depositional landforms.

Table 37 summarises the main depositional landforms.

Table 37 Depositional landforms

Landform	Processes
Spit	Sand or shingle beach ridge extending beyond a turn in the coastline, usually greater than 30°. At the turn, the longshore drift current spreads out and loses energy, leading to deposition. The length of a spit is determined by the existence of secondary currents causing erosion, either the flow of a river or wave action which limits its length
Bayhead beach	Waves break at 90° to the shoreline and move sediment into a bay, where a beach forms. Through wave refraction, erosion is concentrated at headlands and the bay is an area of deposition

Landform	Processes
Tombolo	A sand or shingle bar that links the coastline to an offshore island. Tombolos form as a result of wave refraction around an offshore island which creates an area of calm water and deposition between the island and the coast. Opposing longshore currents may play a role, in which case the depositional feature is similar to a spit
Barrier beach/bar	A sand or shingle beach connecting two areas of land, with a shallow-water lagoon behind. These features form when a spit grows so long that it extends across a bay, closing it off
Hooked/recurved spit	A spit whose end is curved landward, into a bay or inlet. The seaward (distal) end of the spit naturally curves landward into shallower water and the 'hook' may be made more pronounced by waves from a secondary direction to the prevailing wind
Cuspate foreland	Roughly triangular-shaped features extending out from a shoreline. There is some debate about their formation, but one hypothesis suggests they result from the growth of two spits from opposing longshore drift directions

Vegetation plays a very important role in stabilising depositional landforms. Plant succession, in the form of salt marshes and sand dunes, binds the loose sediment together and encourages further deposition.

The sediment cell model

Coastlines operate as sediment systems consisting of three interlinked components (Figure 43 on page 68).

1 Sources: places where sediment is generated, such as cliffs or eroding sand dunes. Some sources are offshore bars and river systems and these are an important source of sediment for the coast.

2 Transfer zones: places where sediment is moving alongshore through longshore drift and offshore currents. Beaches and parts of dunes and salt marshes perform this function.

3 Sinks: locations where the dominant process is deposition and depositional landforms are created, including spits and offshore bars.

Sediment cells have inputs, transfers and outputs of sediment. Under natural conditions the systems operate in a state of dynamic equilibrium, with sediment inputs balancing outputs to sinks. For short periods of time — for instance during a major storm that erodes a spit — the system's equilibrium might be disrupted but it will tend to return to balance over time. Negative feedback mechanisms help maintain the balance by pushing the system back towards balance:

- during a major erosion event a large amount of cliff collapse may occur, but the rock debris at the base of the cliff will slow down erosion by protecting the cliff base from wave attack
- major erosion of sand dunes could lead to excessive deposition offshore, creating an offshore bar that reduces wave energy allowing the dunes time to recover.

Exam tip

Most exam questions about coastal landforms will demand an explanation of how they formed, not a description of them.

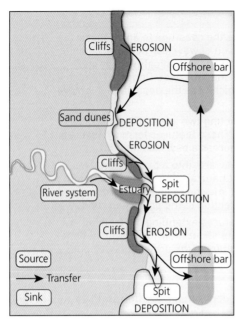

Figure 43 The sediment cell model

Subaerial processes

Weathering

Both **weathering** and mass movement are examples of subaerial processes that are important in the development of coastal landforms.

There are three types of weathering important in sediment production.

1 Mechanical weathering breaks down rocks by the exertion of a physical force and does not involve any chemical change.
2 Chemical weathering involves a chemical reaction and the generation of new chemical compounds.
3 Biological weathering often speeds up mechanical or chemical weathering through the action of plants, bacteria or animals.

The **lithology** of coastal rocks is also a major influence, as some rocks are more prone to some types of weathering than others, as outlined in Table 38.

Table 38 Weathering processes

Type	Process	Explanation	Vulnerable rocks
Mechanical	Freeze–thaw	Water expands by 9% in volume when it freezes, exerting a force within cracks and fissures; repeated cycles force cracks open and loosen rocks	Any rocks with cracks and fissures, especially high on cliffs away from salt spray Freezing is relatively uncommon on UK coasts
	Salt crystallisation	The growth of salt crystals in cracks and pore spaces can exert a breaking force, although less than for freeze–thaw	Porous and fractured rocks, e.g. sandstone The effect is greater in hot, dry climates, promoting the evaporation and the precipitation of salt crystals

Type	Process	Explanation	Vulnerable rocks
Chemical	Carbonation	The slow dissolution of limestone due to rainfall (weak carbonic acid, pH 5.6) producing calcium biocarbonate in solution	Limestone and other carbonate rocks
	Hydrolysis	The breakdown of minerals to form new clay minerals, plus materials in solution, due to the effect of water and dissolved carbon dioxide	Igneous and metamorphic rocks containing feldspar and other silicate minerals
	Oxidation	The addition of oxygen to minerals, especially iron compounds, which produces iron oxides and increases volume contributing to mechanical breakdown	Sandstones, siltstones and shales often contain iron compounds which can be oxidised
Biological	Plant roots	Trees and plants roots growing in cracks and fissures forcing rocks apart	An important process on vegetated cliff tops which can contribute to rock falls
	Rock boring	There are many species of clams and molluscs that bore into rock, and may also secrete chemicals that dissolve rocks	Sedimentary rocks, especially carbonate rocks (limestone) located in the inter-tidal zone

Weathering contributes to rates of coastal recession in a number of ways. Weathering weakens rocks, making them more vulnerable to erosion or mass movement processes. Some strata may be more vulnerable to weathering than others, contributing to the formation of wave-cut notches and therefore having an effect on overall cliff stability. Rates of weathering are very slow. Even in a hot wet climate, basalt, an igneous rock, weathers at a rate of 1–2 mm every 1000 years.

Mass movement

Weathering and erosion are important on many coasts. On some coastlines mass movement is the dominant cause of cliff collapse. There are many different types of mass movement (Table 39), and in a number of types the role of water is very important. Mass movements can be classified in a number of ways, including how rapid the movement is and the type of material (solid rock, debris or soil).

Exam tip

The terms 'landslide' and 'mass movement' are both umbrella terms for a number of more specific processes.

Knowledge check 46

Name two types of chemical weathering.

Mass movement refers to the downslope movement of rock and soil. It is an umbrella term for a wide range of specific movements including landslide, rockfall and rotational slide.

Table 39 Mass movement processes

Fall		Rockfalls, or blockfalls, are a rapid form of mass movement On coasts, blocks of rock can be dislodged by mechanical weathering, or by hydraulic action erosion Undercutting of cliffs by the creation of wave-cut notches can lead to large falls and talus scree slopes at their base
Topple		Geological structure influences topples Where rock strata have a very steep seaward dip, undercutting by erosion will quickly lead to instability and blocks of material toppling seaward
Rotational slide/ slumping		Mass movements can occur along a curved failure surface In the case of a rotational slide, huge masses of material can slowly rotate downslope over periods lasting from days to years Water plays an important role in rotational slides Rotational slides create a back-scar and terraced cliff profile
Flow		Flows are common in weak rocks such as clay or unconsolidated sands These materials can become saturated, lose their cohesion and flow downslope Heavy rainfall combined with high waves and tides can contribute to saturation

How do coastal erosion and sea level change alter the physical characteristics of coastlines and increase risks?

■ Short- and longer-term sea level changes influence the physical geography of coastlines and increase risks for people.

■ Rapid coastal recession happens in some locations because of both physical and human influences.

■ Coastal flooding is a significant risk on some coastlines, worsened by global warming but uncertain in terms of magnitude.

Sea level change

Sea level change is complex because both land level (isostatic) and water level (eustatic) can change over time.

■ A rise or fall in water level causes a eustatic change, which is a global change in the volume of sea water in all the world's seas and oceans.

■ Isostatic change is a local rise or fall in land level.

On a coastline, isostatic and eustatic change can happen at the same time. Table 40 summarises the possible changes.

Table 40 Eustatic and isostatic sea level change

Eustatic fall in sea level	Eustatic rise in sea level
During glacial periods, when ice sheets form on land in high latitudes, water evaporated from the sea is locked up on land as ice, leading to a global fall in sea level	At the end of a glacial period, melting ice sheets return water to the sea and sea level rises globally. Global temperature increases cause the volume of ocean water to increase (thermal expansion) leading to sea level rise
Isostatic fall in sea level	**Isostatic rise in sea level**
During the build up of land-based ice sheets, the colossal weight of ice causes the Earth's crust to sag. When the ice sheets melt, the land surface slowly rebounds upward over thousands of years	Land can 'sink' at the coast because of the deposition of sediment, especially in large river deltas where the weight of sediment deposition leads to very slow 'crustal sag' and delta subsidence

Since the end of the last ice age 12,000 years ago the UK has felt the impact of continuing sea level change:

- Scotland is still rebounding upward, in some places by up to 1.5 mm per year, because of post-glacial adjustment
- in contrast, England and Wales are subsiding at up to 1 mm per year
- the UK is 'pivoting', with the south sinking and the north rising
- sea level rise caused by global warming (eustatic) compounds the effect in the south but reduces it in the north.

Emergent coastlines

The extent of isostatic and eustatic changes during and after the last ice age was large:

- global sea levels fell by 120 m as ice sheets grew
- an equal sea level rise happened over about 1000 years when the ice sheets melted
- in North America and northern Europe the post-glacial isostatic adjustment was up to 300 m.

These two linked changes happened at very different rates:

- post-glacial sea level rise was very rapid, submerging coastlines
- isostatic adjustment was very slow, with land gradually rising out of the sea.

The effect of these changes is to produce an emergent coast, with landforms reflecting previous sea levels. This is shown in Figure 44.

Post-glacial adjustment (sometimes called post-glacial rebound, or post-glacial re-adjustment) refers to the uplift experienced by land following the removal of the weight of ice sheets.

Exam tip

Make sure you use the terms eustatic and isostatic correctly as they are easily confused.

Knowledge check 47

Which leads to a global change in sea level, eustatic or isostatic change?

Exam tip

It is always useful to know located examples of coastlines that are emergent or submergent.

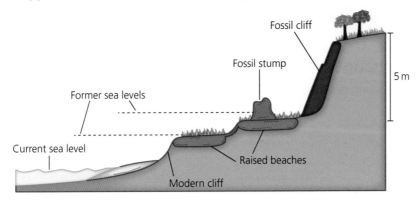

Figure 44 Landforms of an emergent coastline

Submergent coastlines

Coastlines that were never affected by glacial ice cover do not experience post-glacial adjustment. Instead they were submerged (drowned) when post-glacial sea level rose. Submergent coasts are found in southern England and on the east coast of America. The most common coastal landform is a **ria**.

Fjords are submergent landforms found on the coasts of Norway and Canada. Fjords are drowned valleys, but they differ from rias in that:

- the drowned valley is a U-shaped glacial valley
- the fjord is often deeper than the adjacent sea, some over 1000 m deep
- at the seaward end of the fjord there is a submerged 'lip', representing the former extent of the glacier that filled the valley.

The east coast of the USA has **barrier island** landforms (Figure 45):

- they may have formed as lines of coastal sand dunes attached to the shore
- later sea level rise flooded the land behind the dunes forming a lagoon, but the dunes themselves were not eroded and so became islands
- as sea level continued to rise, the dune systems slowly migrated landward.

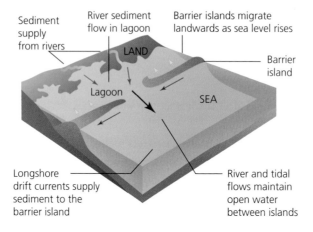

Figure 45 The formation of barrier islands

The Dalmatian Coast consists of limestone anticlines and synclines forming a distinctive coastal landscape of parallel hilly islands (anticlines) and elongated bays (synclines). Post-glacial sea level rise submerged this area, creating the elongated bays in places that were once low-lying valleys.

Contemporary sea level change

The rate of sea level rise today is about 2 mm per year. Figure 46 shows past and projected sea level change.

- Sea level was stable between 1800 and 1870.
- Sea levels rose slowly between 1870 and 1940, but accelerated after that.
- Since 1980, sea level rise has been faster still.
- Between 1870 and today sea level measurements have become more accurate as tide gauges and satellite measurements have become more precise.

> **Rias** are drowned river valleys in unglaciated areas, caused by sea level rise flooding up the river valley, making it much wider than would be expected based on the river flowing into it.

> **Barrier islands** are offshore sediment bars, usually sand-dune covered but, unlike spits, they are not attached to the coast. They are found between 500 m and 30 km offshore and can be tens of kilometres long.

Knowledge check 48

Name an example of a landform that could be called a 'drowned valley'.

Future projections published by the Intergovernmental Panel on Climate Change (**IPCC**) in 2013 range from 28 cm to 98 cm. Some scientists expect global sea level to increase by over 100 cm by 2100.

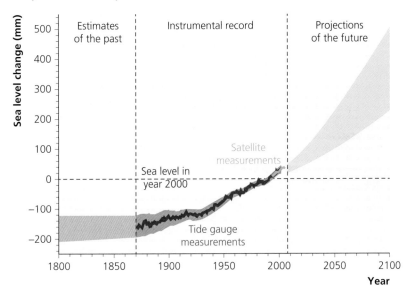

Figure 46 Past and projected sea levels

Sea level is very difficult to predict because there are many factors involved.

- **Thermal expansion** of the oceans depends on how high global temperatures climb.
- Melting of mountain glaciers in the Alps, Himalaya and other mountain ranges will increase ocean water volume.
- Melting of major ice sheets (Greenland, Antarctica) could dramatically increase global sea level.

There is a lot of uncertainty over the contribution of each of these three factors. In addition, sea level can change locally because of tectonic forces. Major earthquakes can force land up or down, sometimes by as much as 2 m.

Rapid coastal retreat

Rapidly eroding coastlines have the following physical features in common:
- long wave fetch, large destructive ocean waves
- soft geology
- cliffs with structural weaknesses such as seaward rock dip and faults
- cliffs which are vulnerable to mass movement and weathering, as well as marine erosion
- strong longshore drift, so eroded debris is quickly removed exposing the cliff base to further erosion.

Human actions can make the situation worse and usually involve interfering with the coastal sediment cell (see Figure 43, page 68). This can happen in a number of ways. The construction of major dams on rivers can trap river sediment behind the dam wall. This then starves the coast of a sediment source, leading to serious consequences.

The **IPCC**, part of the United Nations, is a committee of scientists who periodically review the evidence for global warming.

Exam tip

Learn some data about sea level in the past and projected rises in the future. Make sure you use the correct units, e.g. mm, cm or m.

Thermal expansion, the main driver of sea level rise, occurs because the volume of ocean water increases as global temperatures rise.

Knowledge check 49

By how much is sea level expected to rise by 2100 as a result of global warming?

- The construction of the Aswan High Dam on the River Nile in 1964 reduced sediment volume from 130 million tonnes to about 15 million tonnes per year. Erosion rates jumped from 20–25 metres per year to over 200 metres per year as the delta was starved of sediment.
- The construction of the Akosombo Dam in Ghana in 1965 reduced the flow of sediment down the River Volta from 70 million cubic metres per year to less than 7 million, with major impacts on longshore drift and coastal erosion in Ghana and even in neighbouring countries (Figure 47).

Exam tip

In exam questions, use examples to back up your explanations even if the question is worth only 3–4 marks.

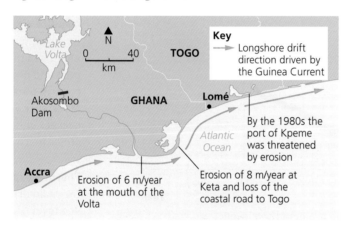

Figure 47 Impacts of the Akosombo dam on the coastal sediment cell

Dredging is another cause of problems. Sand and gravel are often dredged from the sea bed or rivers for use in construction, i.e. to make concrete or foundations, or to make navigation channels deeper. This removes a sediment source, which can have knock-on effects further along a coast by increasing erosion.

Dredging involves scooping or sucking sediment up from the sea bed or a river bed.

Variations in erosion rate

Rates of recession are not constant and are influenced by different factors (wind direction and fetch, seasonal changes to weather systems and the occurrence of storms). On the Holderness Coast in East Yorkshire average annual erosion is around 1.25 m per year but, as Figure 48 shows, there are wide variations in this rate, from 0 m per year to 6 m per year. This is because:

- coastal defences at Hornsea, Mappleton and Withernsea have stopped erosion
- these defences have starved places further south of sediment as groynes have interrupted longshore drift
- erosion rate therefore generally increases from north to south
- some areas of boulder clay are more vulnerable to erosion than others
- some cliffs are more susceptible to mass movement.

Erosion of Holderness varies over time.

- During winter, 2–6 m of erosion is common when storms, combined with spring tides, increase erosion rates.
- Summer erosion, during periods when constructive waves dominate, is much lower.
- Northeasterly storms cause most erosion, because of the long wave fetch of 1500 km from the north Norwegian coast.

The shape of the beach on Holderness can change and promote erosion. Ords are deep beach hollows parallel to the cliff that concentrate erosion in particular locations by allowing waves to directly attack the cliff with little energy **dissipation**. Ords slowly migrate downdrift by about 500 m per year so the location of most erosion changes over time. Ord locations erode four times faster than locations without ords.

Dissipation is the term used to describe how the energy of waves is decreased by friction with beach material during the wave swash up the beach. A wide beach slows waves down and saps their energy so when they break, most energy has gone.

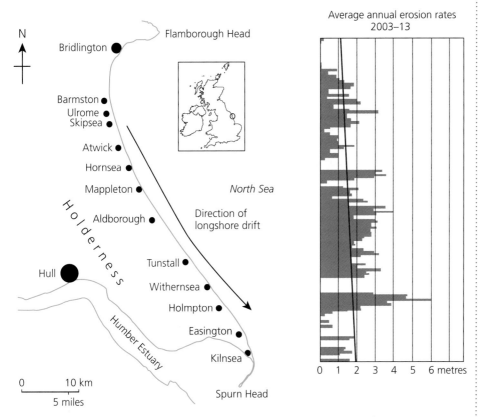

Figure 48 Variations in erosion rate on the Holderness Coast

Exam tip

You may have visited a coastline and done fieldwork there. Learn some key facts about its geology and erosion rates.

Coastal flooding risk

Coastal flooding risk is more widespread than rapid erosion risk. Low-lying coastal land the world over is densely populated for several reasons.

- Coastlines are popular with tourists, especially when access to beaches and the sea is easy.
- Deltas and estuaries are ideal locations for trade between up-river places and places along the coast or across the sea.
- Deltas and coastal plains are especially fertile and ideal for farming.

Many of the world's major river deltas, barely a few metres above sea level, are home to some of the world's largest cities (Table 41). Coastal flooding risk in these river delta locations is made worse by a complex set of processes that increase risk (Figure 49).

Table 41 River delta megacities in Asia

Huang He- Hai	Yangtze	Pearl	Chao Phraya	Ganges-Brahmaputra	Indus
Tianjin (15 million)	Shanghai (24 million)	Guangzhou (11 million)	Bangkok (14.5 million)	Dhaka (14.3 million)	Karachi (23.5 million)

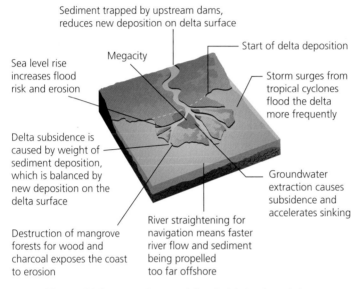

Figure 49 Causes of coastal flood risk in river deltas

In addition to low-lying coastal plains and deltas being at risk from sea level rise, many islands are also at risk.

- In the Indian Ocean the Maldives has a population of 340,000 people spread out across 1200 islands. The highest point in the country is 2.3 m above sea level.
- Sea level rise of 50 cm by 2100 would see 77% of the Maldives disappear into the sea.

Storm surges

The most common cause of coastal flooding is a **storm surge** event caused by:

- a depression (low pressure weather system) in the mid-latitudes, e.g. the UK
- a tropical cyclone (hurricane, typhoon) in areas just north and south of the equator.

Strong winds from these weather systems push waves onshore, increasing the effective height of the sea. If high tides occur at the same time as the storm surge and onshore winds, sea levels increase even more.

Coastal topography can also have an effect. In the North Sea the coastline narrows into a funnel shape for a storm approaching from the north. The storm surge can be funnelled into an increasingly narrow space between coastlines and, as the sea shallows towards the coast, the effect is severe coastal flooding.

The North Sea storm surge of December 2013 was caused by a depression named *Xaver*. It had major impacts because of 80+ mph winds and a storm surge approaching 6 m.

- Significant coastal flooding occurred in Boston, Hull, Skegness, Rhyl and Whitby.
- Scotland's rail network was shut down and 100,000 homes lost power.
- About 2500 coastal homes and businesses in the UK were flooded.
- At Hemsby in Norfolk erosion resulted in several properties collapsing into the sea.
- Across the countries affected, there were 15 deaths.

Forecasting, warning and evacuation, and improved coastal flood defences limited the scale of the damage.

Bangladesh is especially vulnerable to the impacts of tropical cyclone storm surges for a number of reasons:

- much of the country is a low-lying river delta, only 1–3 m above sea level
- incoming storm surges meet out-flowing river discharge from the Ganges and Brahmaputra rivers, meaning river flooding and coastal flooding combine
- intense rainfall from tropical cyclones contributes to flooding
- much of the coastline consists of unconsolidated delta sediment, which is very susceptible to erosion
- deforestation of coastal mangrove forests has removed vegetation that once stabilised coastal swamps and dissipated wave energy during tropical cyclones
- the triangular shape of the Bay of Bengal concentrates a cyclone storm surge as it moves north, increasing its height when it makes landfall.

Three major cyclones have struck Bangladesh since 1970 (Table 42). Death tolls have fallen over time because of much improved warnings, the construction of cyclone shelters and better aid response. Vast areas are flooded by these storm surges, forcing millions of people from their homes and farms in the densely populated coastal areas.

Table 42 Bangladesh cyclones

	Storm surge height	Maximum 1-minute sustained winds speed	Lowest air pressure	Deaths and economic losses
1970 Bhola Cyclone	10 m	205 km/h	966 mb	300,000–500,000 US$90 million
1991 Cyclone	6 m	250 km/h	918 mb	139,000 US$1.7 billion
2007 Cyclone Sidr	3 m	260 km/h	944 mb	15,000 US$1.7 billion

Climate change and coastal flood risk

The global scientific authority on the link between global warming and coastal flood risk is the IPCC. Its 5th Assessment Report, published in 2014, came to the conclusions shown in Table 43.

Table 43 Summary of coastal risks from global warming (IPCC, 2014)

Sea level	Delta flooding	Tropical cyclones
Sea level will rise by between 28 cm and 98 cm by 2100, with the most likely rise about 55 cm by 2100	The area of the world's major deltas at risk from coastal flooding is likely to increase by 50%	The frequency of tropical cyclones is likely to remain unchanged, but there could be more of the largest storms
Storm surges	**Wind and waves**	**Coastal erosion**
Storm surges linked to depressions are likely to become more common	There is some evidence of increased winds speeds and large waves	Erosion will generally increase because of the combined effects of changes to weather systems and sea level

The IPPC report shows that some risks are much more certain than others. The magnitude and timing of these changes is very uncertain.

How can coastlines be managed to meet the needs of all players?

- Erosion, flooding and sea level rise have consequences for people and their property.
- A wide range of management strategies and engineering approaches can be used to reduce risk, but these have advantages and disadvantages.
- Decision-making about coastal management can lead to conflict as well as winners and losers.

Consequences of coastal recession on communities

The costs of rapid coastal recession can be classified into three broad categories (Table 44).

Table 44 Costs of coastal recession

Economic costs	Social costs	Environmental costs
Loss of property in the form of homes, businesses and farmland. These are relatively easy to quantify	Costs of relocation and loss of livelihood/jobs (which can be quantified) but also include impact on health (such as stress and worry) which are much harder to quantify	Loss of coastal ecosystems and habitats. These are almost impossible to quantify financially but are likely to be small

Losses because of erosion tend to be very localised and costs are very specific to those locations. In 2015 farmland in the UK had an average value of about £20,000 per hectare, whereas residential land can vary from £500,000 to £2.5 million per hectare. If roads are lost to erosion they can cost between £150,000 and £250,000 per 100 m to re-route and replace. There are also losses in terms of amenity value and economic losses to businesses if coastlines become unattractive and depopulated.

Economic losses because of erosion are small because:

- erosion happens slowly with a small number of properties affected over a long period of time
- property that is at risk loses its value long before it is destroyed by erosion, because potential buyers recognise the risk
- areas of high-density population, especially towns and villages, tend to be protected by coastal defences.

However, for those communities affected the losses can be significant if, for instance, a whole village is at risk. For coastal people erosion means:

- falling property values, as the date of eventual loss approaches
- an inability to sell their property because the possibility of loss by erosion is too great
- an inability to insure against the loss (coastal erosion is not covered)
- the loss of their major asset, and facing the costs of getting a new home
- an increasingly unattractive environment scarred by collapsing cliffs, failing sea defences and blocked roads and paths.

There is very little help available to people about to lose their homes to the sea. In the UK it consists of 'Coastal Change Pathfinder' projects, which:

- cover the cost of property demolition and site restoration
- provide up to £1000 in relocation expenses such as removal vans and storage
- provide up to £200 in hardship expenses
- have 'rollback' policies, giving people fast-tracked planning approval to build a new home somewhere else.

> **Exam tip**
>
> Be careful not to over-exaggerate erosion risk: few people in the UK are threatened by it directly, many more are at risk from coastal flooding.

Consequences of coastal flooding on communities

Coastal floods and storm surges are one-off events that occur perhaps decades apart, whereas erosion is a continual process. Flooding tends to be larger in areal extent and involve greater losses, i.e. in some cases it can be classified as a natural disaster (Table 45).

Table 45 The social and economic impacts of coastal storm surges in developed and developing countries

Example	Cause	Economic costs	Social costs
Netherlands 1953 (North Sea flood)	Mid-latitude depression moving south through the North Sea generating a 5 m storm surge	Almost 10% of Dutch farmland flooded 40,000 buildings damaged and 10,000 destroyed	1800 deaths
UK 2013–2014 Winter storms	Coastal and other flooding caused by a succession of depressions and their storm surges	Damage of around £1 billion over the course of the winter	17 deaths (from all causes)
USA 2012 Hurricane Sandy	Landfall of hurricane Sandy in New Jersey and other US states with a storm surge up to 4 m	US$70 billion in damage 6 million people lost power and 350,000 homes in New Jersey were damaged or destroyed	71 deaths
Philippines 2013 Typhoon Haiyan	One of the most powerful tropical storms ever with a 4–5 m storm surge	Damages of around US$2 billion, centred on the city of Tacloban	At least 6,300 deaths, 30,000 injured

Environmental refugees

By 2100, in some places sea level rise resulting from global warming will be very difficult to manage. The most at-risk places are low-lying islands including the Maldives, Tuvalu, the Seychelles and Barbados. These small islands have particular risk factors.

- Tuvalu's highest point is 4.5 m above sea level, and most land is 1–2 m above sea level.
- 80% of people in the Seychelles live and work at the coast.
- Coral reefs, which act as a natural coastal defence against erosion, are being destroyed by global warming-induced coral bleaching.
- Water supply is limited and at risk from salt-water incursion as sea level rises and groundwater is over-used.
- They have small and narrow economies based on tourism and fishing, which are easily disrupted.
- They have high population densities and very limited space, so no opportunity for relocation.

The worst-case scenario for Tuvalu, and part of the Maldives, is that some or all islands will have to be abandoned, creating **environmental refugees**.

Managing coastal recession and flood risk

Hard engineering

The traditional management approach for coastal erosion and/or flooding is to encase the coastline in concrete, stone and steel. This aims to directly stop physical processes altogether (such as erosion or mass movement) or alter them to protect the coast (such as encouraging deposition to build larger beaches). This approach has a number of advantages and disadvantages (Table 46).

Table 46 Advantages and disadvantages of hard engineering

Advantages	Disadvantages
■ Obvious to at-risk people that 'something is being done' to protect them ■ A 'one-off' solution that could protect a stretch of coast for decades	■ Costs are usually very high, and there are on-going maintenance costs ■ Even very carefully designed engineering solutions are prone to failure ■ Coastlines are made visually unattractive and the needs of coastal ecosystems are usually overlooked ■ Defences built in one place frequently have adverse effects further along the coast

The economic costs of hard engineering are very high. Groynes cost £150–250 per metre, sea walls £3000–10,000 and rip-rap £1300–6000. Table 47 summarises the most common types of hard engineering.

Knowledge check 52

Which is more likely to cause widespread economic losses and even deaths, coastal erosion or coastal flooding?

Environmental refugees are communities forced to abandon their homes because of natural processes, including sudden ones such as landslides or gradual ones such as erosion or rising sea levels.

Exam tip

It is easy to get bogged down in describing types of coastal defence rather than explaining their purpose or assessing their effectiveness.

Table 47 Hard engineering coastal defences

Type	Construction and materials	Purpose	Impact on physical processes
Rip-rap (rock armour)	Large igneous or metamorphic rock boulders, weighing several tonnes	Break up and dissipate wave energy Often used at the base of sea walls to protect them from undercutting and scour	Reduced wave energy Sediment deposition between rocks May become vegetated over time
Offshore rock breakwater	Large igneous or metamorphic rock boulders weighing several tonnes	Forces waves to break offshore, rather than at the coast, reducing wave energy and erosive force	Deposition encouraged between breakwater and beach Can interfere with longshore drift
Sea wall	Concrete with steel reinforcement and deep piled foundations; can have a stepped and/or 'bullnose' profile	A physical barrier against erosion They often also act as flood barriers Modern sea walls are designed to dissipate, not reflect, wave energy	Destruction of the natural cliff face and foreshore environment If reflective, can reduce beach volume
Revetments	Stone, timber or interlocking concrete sloping structures, which are permeable	To absorb wave energy and reduce swash distance by encouraging infiltration Reduce erosion on dune faces and mud banks	Reduced wave power Can encourage deposition and may become vegetated

Type	Construction and materials	Purpose	Impact on physical processes
Groynes **Plan view:**	Vertical stone or timber 'fences' built at 90° to the coast, spaced along the beach	To prevent longshore movement of sediment, and encourage deposition building a wider, higher beach	Deposition and beach accretion Prevention of longshore drift, sediment starvation and increased erosion downdrift

Soft engineering

Soft engineering is an alternative to hard engineering that attempts to work with natural physical systems and processes to reduce the coastal erosion and flood threat. Soft engineering is usually less obvious and intrusive at the coast, and may be cheaper in the long term. However, it is not suitable for all coasts (Table 48).

Knowledge check 53

Which type of hard engineering interferes most with longshore drift?

Table 48 Soft engineering methods

Soft engineering method	Technique	Cost and issues
Beach nourishment	Artificial replenishment of beach sediment to replace sediment lost by erosion, to enlarge the beach so that it dissipates wave energy and reduces wave erosion and increases the amenity value of the beach	Costs of £2 million per km of beach are typical, but ongoing costs are high and sediment must not be sourced from elsewhere in the local sediment cell
Cliff regrading and drainage	Cliff slope angles reduced to increase stability and revegetated to reduce surface erosion. In-cliff drainage reduces pore-water pressure and mass movement risk	Costs of £1 million per 100 m are common. Can be disruptive during construction
Dune stabilisation	Fences are used to reduce wind speeds across the dunes, which are then replanted with marram and lyme grass to stabilise the surface. This reduces erosion by wind and water	Dune fencing costs £400–2000 per 100 m and replanting dunes about £1000 per 100 m. This means that working to maintain natural sand dunes can be very cost-effective in the long term

Sustainable coastal management

Coastal communities around the world face the dynamic nature of the coast's every-day environment. They increasingly face threat from:

- rising global sea levels, but there is uncertainty about the scale and timing of the rise
- increased frequency of storms and the possibility of increased erosion and flooding.

To cope with these threats, communities need to adapt and employ sustainable coastal management to ensure the wellbeing of people and the coastal environment (Figure 50).

Sustainable coastal management means managing the wider coastal zone in terms of people and their economic livelihood, social and cultural wellbeing, safety from coastal hazards, as well as minimising environmental and ecological impacts.

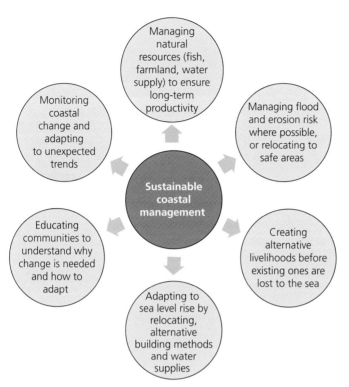

Figure 50 Aspects of sustainable coastal management

Adopting sustainable coastal management may lead to **conflict** because:

■ coastal natural resources may have to be used less in order to protect them — so some people lose income

■ relocation may be needed where engineering solutions are too costly or not technically feasible

■ some erosion and/or flooding will always occur, as engineering schemes cannot protect against all threats

■ future trends, such as sea level rise, may change, creating uncertainty and the need to change plan.

Integrated Coastal Zone Management (ICZM)

ICZM is a holistic approach increasingly used to manage coasts. It dates from the Rio Earth Summit in 1992 and has a number of key characteristics.

1 The entire coastal zone is managed, not just the narrow zone where breaking waves cause erosion or flooding. This includes all ecosystems, resources and human activity in the zone.

2 It recognises the importance of the coastal zone to people's livelihoods as, globally, very large numbers of people live and work at the coast — but their activities tend to degrade the coastal environment.

3 It recognises that management of the coast must be sustainable, meaning that economic development has to take place to improve the quality of life of people but that this needs to be environmentally appropriate and equitable (benefit everyone).

Exam tip

Take care when using the word 'conflict'. In a coastal context it means 'disagreement' rather than 'armed conflict'!

Conflict, in the context of coastal management, means disagreement over how the coast should be protected from threats and which areas should be protected. Often conflict exists between different stakeholders, such as residents versus the local council.

ICZM is coastal management planning over the long term, involving all stakeholders, working with natural processes and using 'adaptive management', i.e. changing plans as threats change.

ICZM works with the concept of littoral cells or sediment cells. The coastline can be divided up into littoral cells and each cell managed as an integrated unit.

■ In England and Wales there are 11 sediment cells.
■ Each cell is managed either as a whole unit or a sub-unit.
■ In both cases a plan called a Shoreline Management Plan (SMP) is used.
■ The SMP area is further divided into sub-cells.
■ SMPs extend across council boundaries, so many councils must work together on an agreed SMP to manage an extended stretch of coastline.

Policy decisions

In the UK, coastal management is overseen by DEFRA (the government Department for Environment, Food and Rural Affairs). Since DEFRA introduced Shoreline Management Plans in 1995 there have been only four policies available for coastal management. These are very different in terms of their costs and consequences (Table 49).

Table 49 Coastal management policy options

No active intervention	Hold the line
No investment in defending against flooding or erosion, whether or not coastal defences have existed previously. The coast is allowed to erode landward and/or flood	Build or maintain coastal defences so that the position of the shoreline remains the same over time
Strategic (managed) realignment	**Advance the line**
Allow the coastline to move naturally (in most cases to recede) but managing the process to direct it in certain areas. Sometimes called 'strategic realignment' or 'strategic retreat'	Build new coastal defences on the seaward side of the existing coastline. Usually this involves land reclamation

Making decisions about which policy to apply to a particular location is complex. It depends on a number of different factors including:

■ the economic value of the assets that could be protected, e.g. land
■ the technical feasibility of engineering solutions: it may not be possible to 'hold the line' for mobile depositional features such as spits, or very unstable cliffs
■ the cultural and ecological value of land: it may be desirable to protect historic sites and areas of unusual biodiversity
■ pressure from communities: vocal local political campaigning to get an area protected
■ the social value of communities that have existed for centuries.

SMPs plan for the future using three time periods, called 'epochs'. These are: up to 2025, 2025–2055, and beyond 2055. A hold-the-line policy applied to an area up to 2025 may become a managed realignment policy after 2025. This is because by 2025 sea level rise is likely to have made 'hold the line' a much more expensive policy to apply.

Cost–benefit analysis

Cost–benefit analysis (CBA) is used to help decide if defending a coastline from erosion and/or flooding is economically justifiable. An example is Happisburgh in

Littoral cells contain sediment sources, transport paths and sinks. Each littoral cell is isolated from adjacent cells, and can be managed as a holistic unit.

Knowledge check 54

What is the name given to a managed unit of coastline that has its own plan, often extending across council boundaries?

Exam tip

Some acronyms, such as ICZM and SMP, are so well-known that you can use them in the exam without writing them out in full.

Knowledge check 55

Which coastal management policy option involves 'letting nature take its course'?

North Norfolk. The policy adopted in this area is 'no active intervention'. This is because to defend the village would have an impact on the wider coastal management plan. Happisburgh would end up as a promontory, blocking longshore drift and causing further erosion downdrift. Longer term, the plan is managed realignment, although this would still involve property being lost to the sea by erosion. Some of the costs and benefits are shown in Table 50.

Table 50 Cost–benefit analysis for Happisburgh

Hold-the-line costs for a 600 m stretch of coastline: ■ seawall: £1.8–6 million ■ rip-rap: £0.8–3.6 million ■ groynes: £0.1–1.5 million	
Costs of erosion	**Benefits of protection**
■ £160,000 could be available to the Manor Caravan Park to assist in relocating to a new site ■ Affected residents could get up to £2000 each (a total cost of £40,000–70,000) in relocation expenses plus the cost to the council of finding plots of land to build new houses ■ Grade 1 listed St Mary's church and Grade 2-listed Manor House would be lost ■ Social costs as the village is slowly degraded, including health effects and loss of jobs	■ By 2105, between 20 and 35 houses would be 'saved' from erosion with a combined value of £4 million–£7 million (average house price £200,000 in 2015) ■ 45 hectares of farmland would be saved, with a value of £945,000 ■ The Manor Caravan Park would be saved, which employs local people

The cost of building coastal defences at Happisburgh is around £6 million. This is very close to the value of property that could be saved, and much higher than the compensation costs payable to local residents. Coastal managers argue that Happisburgh must be seen in the wider context of the whole SMP, further justifying the decision not to defend the village.

Environmental Impact Assessment

Coastal management usually requires an Environmental Impact Assessment (EIA) to be carried out. This is quite separate from any cost–benefit analysis, although it might inform the final CBA. EIA is a process that aims to identify:

■ the short-term impacts of construction on the coastal environment
■ the long-term impacts of building new sea defences or changing a policy from hold the line to no active intervention or managed realignment.

EIA is wide-ranging and includes assessments of:

■ impacts on water movement (hydrology) and sediment flow, which can affect marine ecosystems because of changes in sediment load
■ impacts on water quality, which can affect sensitive marine species
■ possible changes to flora and fauna, including marine plants, fish, shellfish and marine mammals
■ wider environmental impacts such as air quality and noise pollution, mainly during construction.

Conflict, winners and losers

Coastal management decisions directly affect people's lives. These effects can be positive or negative, producing perceived:

Exam tip

A worked cost–benefit analysis, such as the one for Happisburgh, could prove useful in the exam as a case study.

Knowledge check 56

What is the main purpose of a cost–benefit analysis on proposed coastal defences?

- winners: people who gain from a decision, either economically (their property is safe), environmentally (habitats are conserved) or socially (communities can remain in place)
- losers: people who are likely to lose property, their business or job, be forced to move, or see the coastline 'concreted over' and view this as an environmental negative.

In some ways this is inevitable because:

- coastal managers produce plans for entire SMP areas, so some areas are protected but others are not
- local councils and government (DEFRA) have limited resources, meaning all places cannot be protected.

There are examples where all stakeholders agree on a course of action. The Blackwater Estuary in Essex is an area of tidal salt marsh and low-lying farmland. Prone to flooding and coastal erosion, the farmland was traditionally protected by flood embankments and revetments. Over the last 30 years it has become clear that responding to rising sea levels and greater erosion by building more and higher coastal defences in places such as Blackwater is not sustainable.

The solution adopted was radical. In 2000 Essex Wildlife Trust purchased Abbotts Hall Farm on the Blackwater Estuary, which was threatened by erosion and flooding. A 4000 hectare managed realignment scheme was implemented by creating five breaches in the sea wall in 2002. This allowed new salt marshes to form inland. The scheme has a number of benefits.

- The Abbots Hall Farm owners received the market price for their threatened farm.
- The very high costs of a 'hold the line' policy were avoided, but flood risk was reduced.
- Water quality in the estuary improved because of expansion of reed-beds that filter and clean the water.
- New paths and waterways were created for leisure activities.
- Additional income streams from ecotourism and wildlife watching were created.
- Important bird (dunlin, redshank, geese) and fish (bass and herring) nurseries were enhanced.

The Blackwater Estuary shows that environmentalists, landowners, coastal managers and local people and businesses can all be kept happy, even when radical plans are adopted.

Coastal management in the developing world

In many parts of the developing world, such as the Maldives, parts of Vietnam and the West African coast, erosion is rapid, often because of a combination of:

- upstream dams reducing sediment supply to the coast and disrupting local sediment cells
- rapid unplanned coastal development, urbanisation and the development of tourist resorts with piecemeal defences and no overall plan
- widespread destruction of mangrove forests for fuelwood and shrimp-ponds, exposing soft sediments to rapid erosion.

In many cases, the main 'losers' are the poorest people. Farmers and residents usually lack a formal land-title so cannot claim compensation (even if it were available). Coastlines become more vulnerable to sea level rise, the impact of tropical cyclone storm surges and even tsunami. When these disasters strike it is the poorest that lose everything. In many cases it is individual property owners that take responsibility for coastal defences in the absence of local council or government plans.

Summary

- Coastal landscapes and the littoral zone are dynamic, vary in type from rocky to coastal plain and can be high or low energy.
- Geological structure – including concordant and discordant strata, faulting and jointing – is important in shaping the coastal landscape.
- Lithology (igneous, sedimentary, metamorphic, unconsolidated) is a strong influence on rate of coastal recession but vegetation is important in stabilising some coasts.
- Beach profiles vary as a result of constructive and destructive waves, and cliff profiles and landforms are influenced by erosion (abrasion, corrosion, hydraulic action).
- Sediment is transported as part of a sediment cell, which includes landforms produced by deposition and processes such as longshore drift and plant succession.
- Subaerial processes include weathering and mass movement and these are important on some coastlines, especially those with weak geology.
- Sea level change involves a complex interplay of eustatic and isostatic factors and short- and longer-term changes, producing emergent and submergent coastlines with characteristic landforms.
- Some coastlines experience rapid coastal recession resulting from a mix of human and physical factors, while other coasts experience short- and longer-term coastal flood risk.
- Significant local economic losses can result from erosion and/or flooding, with major consequences for vulnerable communities.
- Hard and soft engineering approaches can both be used to manage erosion and flooding, but each has advantages and disadvantages.
- The goal of sustainable coastal management can be achieved through a combination of ICZM, policy decisions, cost–benefit analysis and environmental impact assessment, but the potential for conflict over decisions is high.

Questions & Answers

■ Assessment overview

In this section of the book, two sets of questions on each of the content areas are given, one set for AS and one for A-level. For each of these, the style of questions used in the examination papers has been replicated, with a mixture of short answer questions, data response questions and extended writing questions. At AS there may be a small number of multiple choice questions that test basic recall or ability to select information from a stimulus resource. The relative proportions and weightings of the marks varies between AS and A-level.

All questions that carry a large number of marks (at AS and A-level) require candidates to consider connections between the subject matter and to demonstrate deeper understanding in order to access the highest marks. At **AS** the breakdown of the questions per topic is:

■ Tectonic processes and hazards (28 marks)
  Typical question sequence: 1, 2, 3, 4, 6 and 12 mark questions
■ **And either** Glaciated landscapes and change (28 marks)
  Typical question sequence: 1, 2, 3, 4, 6 and 12 mark questions
■ **Or** Coastal landscapes and change (28 marks)
  Typical question sequence: 1, 2, 3, 4, 6 and 12 mark questions

In the AS Paper 1, there are also fieldwork skills questions on Coastal landscapes and change and Glaciated landscapes and change. The overall paper takes 1 hour and 45 minutes and is worth a total of 90 marks, making up 50% of the AS qualification. Fieldwork skills are covered in Student Guide 4 on Geographical skills. There are no fieldwork skills questions on the A-level exam paper.

At **A-level** the breakdown of the questions per topic is:

■ Tectonic processes and hazards (16 marks)
  Typical question sequence: 4 and 12 mark questions
■ **And either** Glaciated landscapes and change (40 marks)
  Typical question sequence: 6, 6, 8 and 20 mark questions
■ **Or** Coastal landscapes and change (40 marks)
  Typical question sequence: 6, 6, 8 and 20 mark questions

The A-level Paper 1 also includes a question on each of the Water cycle and water insecurity, and The carbon cycle and energy security — topics that are not covered in this guide. Overall, the paper takes 2 hours and 15 minutes and is worth a total of 105 marks, making up 30% of the A-level qualification.

In this section of the book, each of the content areas are structured as follows:

- sample questions in the style of the examination
- Levels-based mark schemes for extended questions (6 marks and over) in the style of the examination
- one student answer per question — at the upper level
- examiner commentary on each of the above.

Carefully study the descriptions given after each question to understand the requirements necessary to achieve a high mark. You should also read the commentary with the mark schemes to understand why credit has or has not been awarded. In all cases, actual marks are indicated.

AS questions

Tectonic processes and hazards

Question 1

(a) Name **one** secondary hazard that can result from volcanic eruptions. (1 mark)

(b) Study Table 1.

Volcano	Country	Year	VEI	Deaths
Nyiragongo	DR Congo	2002	1	47
Chaiten	Chile	2008	4	1
Mount Merapi	Indonesia	2010	4	353
Nabro	Eritrea	2011	4	31
Mount Sinabung	Indonesia	2014	2	15
Kelud	Indonesia	2014	4	2
Mount Ontake	Japan	2014	3	61

(i) Suggest **one** reason for the differences in Volcanic Explosivity Index (VEI) shown. (2 marks)

(ii) Suggest **one** reason why the eruptions with the highest VEI shown in Table 1 did not always cause the most deaths. (3 marks)

e Part 1(a) is a simple recall question. Part 1b(i) requires use of Table 1 to identify that the VEI varies from 1 to 4 and give one reason. This needs to be an extended point for 2 marks (two reasons cannot gain 2 marks). Part 1b(ii) needs a little more explanation because 3 marks need to be gained from one extended point. An example could be used as part of this answer.

Student answer

(a) Lahars

(b) (i) VEI measures how explosive a volcano is, subduction zone volcanoes have high VEI of 3 or 4 because their magma has a high gas and silica content.

(ii) The management of the hazard could explain why high VEI events such as Chaiten and Kelud have low death tolls. If the eruption was predicted, and people evacuated, the death toll would be reduced despite the high VEI, whereas the eruption of Nyiragongo may have been unexpected and in a remote location with no hazard management systems.

🄮 **6/6 marks awarded.** This is a good answer. Part 1(a) scores 1 mark as lahars are a secondary volcanic hazard. In 1(b)(i) the idea of subduction zones is extended with an explanation of explosivity related to magma type (2 marks). Part 1(b)(ii) takes the single idea of management and extends it in relation to prediction/evacuation and a lack of management in place (using an example from Table 1) so scores 3 marks.

(c) Explain **two** reasons why the number of deaths caused by volcanic eruptions has fallen in the last 30 years. (4 marks)

(d) Explain the causes of earthquakes. (6 marks)

🄮 Part 1(c) requires two reasons, each with an extension to gain 2×2 marks. Be careful not to write in detail about only one reason here. Part 1(d) is a more extended question. It is about physical geography processes, so a detailed sequential explanation is needed with good use of terminology to gain full marks. For questions worth 6+ marks, it is always a good idea to use examples to add depth. The Levels mark scheme for a 6-mark question is shown below. The same Levels mark scheme is used for the 6-mark questions in the AS options questions.

Level 1 1–2 marks	▪ Demonstrates isolated elements of geographical knowledge and understanding, some of which may be inaccurate or irrelevant ▪ Understanding addresses a narrow range of geographical ideas, which lack detail
Level 2 3–4 marks	▪ Demonstrates geographical knowledge and understanding, which is mostly relevant and may include some inaccuracies ▪ Understanding addresses a range of geographical ideas, which are not fully detailed and/or developed
Level 3 5–6 marks	▪ Demonstrates accurate and relevant geographical knowledge and understanding throughout ▪ Understanding addresses a broad range of geographical ideas, which are detailed and fully developed

Student A

(c) Firstly, prediction of volcanic eruptions has improved because of the use of gas spectrometers and tiltmeters that can accurately predict eruption timing. Secondly, predictions allow more timely evacuations, which are made possible by hazard management organisations such as Phivolcs in the Philippines, thus reducing death tolls.

(d) Earthquakes are caused along fault lines which are found at plate boundaries such as the San Andreas fault on the conservative plate boundary in California. As the tectonic plates move, they tend to stick rather than move easily. This means that pressure builds up which is released as an earthquake. The earthquake is released energy that radiates out from the point of origin on the fault line called the focus.

ⓔ 9/10 marks awarded. The answer to 1(c) scores 4 marks. There are two clear reasons (although they are linked) and in the case of evacuation, an example is used to add some depth to the answer. 1(d) scores 5 marks. This is a good answer which refers to an example and also most of the sequence of how earthquakes are caused. It could refer to P, S and L waves at the end to add some additional detail.

Landscape systems, processes and change: Glaciated landscapes and change

Question 2

(a) State **one** erosion process that occurs in upland glacial landscapes. (1 mark)

(b) (i) Study Figure 1. Calculate the difference between the minimum and maximum mass balance for the Wolverine glacier. Show your working. (2 marks)

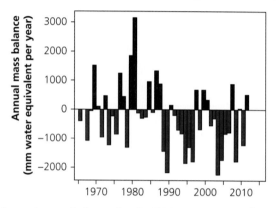

Figure 1 Annual mass balance for the Wolverine glacier (Alaska, USA)

(ii) Explain **one** reason for the changes in mass balance shown on Figure 1. (3 marks)

ⓔ Parts (a) and (b) are a combination of recall (a) and interpretation and explanation of a data stimulus resource. Figure 1 needs to be studied carefully, taking note of the timescale (1965–2012) and any trends that can be seen. There is a shift towards negative mass balances from 1990 onwards. Part (b)(ii) requires an extended explanation of one reason, so be careful not to give three short reasons because this would only gain one mark.

Student answer

(a) Plucking

(b) (i) +3150 in 1981 to because −2200 in 2004 = 5350 mm w.e. per year.

(ii) Recent global warming could be responsible for the shift to negative mass balances since 1990. Higher global temperatures mean ablation exceeds accumulation, leading to negative mass balance. A shorter winter could have led to lower snowfall reducing accumulation on the glacier.

ℯ 6/6 marks awarded. The answer to Part (a) is correct for 1 mark. Abrasion would be an alternative correct answer. The student correctly identifies the maximum and minimum years from Figure 1 for Part (b)(i) and shows these in their answer, as well as working out the difference between them correctly scoring 2 marks. Mark schemes for questions such as this have an 'acceptable' range of correct answers, usually ±5% around the precise answer. Notice that in Part (b)(ii) the basic reason — global warming — is stated, followed by two further points related to this reason, so this answer scored 3 marks.

(c) Explain the formation of **two** fluvioglacial landforms.   (4 marks)

(d) Explain the importance of freeze–thaw weathering in the formation of periglacial landforms.   (6 marks)

ℯ Part (c) is a point-marked question. Take care to choose fluvio-glacial landforms not any glacial landforms. In addition 2 × 2 mark developed explanations are needed rather than a long explanation of only one landform. Part (d) is a Level-marked question (see Levels mark scheme on page 90). Answers need to demonstrate an understanding of frost-shattering and link it to landforms (plural). Questions such as this, which focus on physical processes, need good use of terminology and logical, sequential explanations.

Student answer

(c) Eskers are fluvioglacial landforms which consist of long narrow ridges of sediment. They formed by deposition from subglacial or englacial streams as velocity drops and sediment is deposited and the sediment has some rounding and sorting indicating transport by meltwater. Sandhurs are outwash plains at the snout of a glacier, and are formed by deposition as fast-flowing meltwater slows and sediment is reworked by meltwater channels flowing across the sandhur with continued deposition building up layers of sediment.

(d) Freeze–thaw weathering is one of the most important physical processes in periglacial areas. It relies on repeated freezing–thawing cycles which is a feature of the periglacial climate. Water freezes in cracks and expands by 9% in volume exerting a force that can split rocks into angular fragments. These angular fragments are the basis of patterned ground formation as the related process of frost-heave moves larger freeze–thaw weathered rocks towards the surface to form striped polygons. Frost-heave is the repeated expansion and contraction of the permafrost layer from season to season. Most periglacial landforms, such as solifluction lobes and terraces, and blockfields, consist of freeze–thaw debris.

ℯ 10/10 marks awarded. The answer to Part (c) scores 4 marks. There is no mark for naming the landform or describing it. The esker explanation has two valid explanations: deposition in sub-glacial and englacial streams and the sediment type indicating formation process. The sandur explanation also has two explanations: deposition as meltwater slows and the idea of layers building up.

The answer to Part (d) scores 6 marks. It uses good terminology throughout and explains the process of freeze–thaw weathering sequentially. There is detailed understanding, such as reference to water expanding by 9% on freezing. The process is then linked to the related process of frost-heave and then to a number of named landforms, i.e. patterned ground, solifluction lobes and blockfields. It is a good answer because it recognises that freeze–thaw is one of a number of processes operating in tandem in periglacial areas rather than just considering it in isolation.

(e) **Assess the significance of present-day local and global threats to glacial upland landscapes.**

(12 marks)

e This extended question uses the command word 'assess' which means 'weigh-up'. The key here is to demonstrate understanding of a range of both global and local threats, using examples to support your answer (evidence) and then make a judgement about their significance, i.e. how important or severe are they. Answers must focus on upland glacial landscapes. These could be active or relict, as the question does not specify one type. Level 3 marks are achieved by making supported judgements about the threats. Answers that just explain the threats are likely to get no more than 8 marks. Successful arguments would include ones that argue local threats are less significant than global ones, or that pick a particular threat and make a case for it being the most severe. Questions in this format do not have a 'right' answer; it is the quality of your assessment and judgements that gain marks.

Student answer

(e) Climate change, in the form of global warming, is the main global-scale threat to upland glacial areas and the most significant threat overall **a**. All alpine glaciers are at risk as the enhanced greenhouse effect has increased global temperatures by 1°C since 1900 and risk further increases of 3–4°C by 2100 **b**. Ablation zones have expanded and glacier snouts have retreated up-valley. All of France's 6 Alpine glaciers are retreating. Many Himalayan glaciers are retreating by 20 m per year **b**. This is the biggest threat because it puts at risk the existence of glaciers. Glacial ice on Mt Kilimanjaro in Africa has shrunk by 80% since 1912. Warming also risks alpine ecosystems, and threatens relict glacial landscapes like the Lake District as their ecology and physical environment changes **c**.

Global warming could increase the risk from local threats **d**. Tourism is a major threat in the Alps already as the popularity of skiing rises and urban development encroaches into valleys **e**. Warmer temperatures could increase accessibility of more isolated upland glacial areas such as the Himalaya and Patagonia. A longer summer tourism season could increase visitor numbers leading to more footpath erosion, pressure to improve roads and rail links and develop power sources such as HEP in glacial valleys **e**. However, local threats such as tourism and development

> pressure can be managed, even internationally by strategies such as the Alpine Convention or local approaches such as National Parks ⓘ. These can reduce local risks, but they do nothing to reduce the context threat of global warming. Even the 2015 COP21 UN climate change agreement will not stop global warming but it may reduce its severity ⓖ.

ⓔ **12/12 marks awarded.** This is a Level 3 answer. ⓐ It makes a clear judgement about which threat is the most significant. ⓑ This is supported by some detailed evidence of the threat posed by global warming in several different locations. ⓒ In addition to covering active upland glacial environments the answer also considers relict ones. ⓓ Rather than seeing local and global threats as separate, a link is made between the two which shows very good understanding. ⓔ A number of local threats are considered with some support. ⓕ The argument is supported by the point that local threats are less severe because they can be managed and some examples are provided, whereas ⓖ the fact that global warming is very difficult to manage is also considered.

Landscape systems, processes and change: Coastal landscapes and change

Question 3

(a) State **one** type of coastal ecosystem. (1 mark)

(b) (i) Study Figure 2. Calculate the difference in mean annual erosion rate between Overstrand and Trimingham. Show your working. (2 marks)

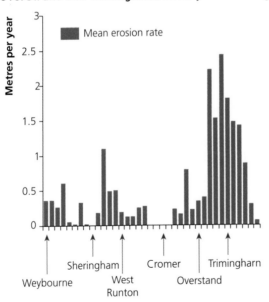

Figure 2 Mean annual erosion rate on a stretch of the Norfolk Coast

(ii) Explain **one** possible reason for the differences in erosion rate shown on Figure 2. (3 marks)

ⓔ Part (a) is a recall question. Part (b)(i) requires some analysis of Figure 2 and includes a calculation. Remember that most question sequences will contain some marks that test numerical skills. Part (b)(ii) should be answered with reference to one reason only. This means writing a sequence of linked, extended points to gain the 3 marks available.

Student answer

(a) Saltmarsh

(b) (i) Overstand = 0.35 m, Trimingham = 1.8 m, 1.8 – 0.35 = 1.45 m

(ii) Different types of coastal protection could explain the differences. Cromer and Sheringham could have sea-walls reducing erosion to 0 m whereas Overstrand may have groynes which don't directly protect the cliff, but could trap sand and prevent longshore drift making erosion worse downdrift in Trimingham.

ⓔ **6/6 marks awarded.** The answer to Part (a) is correct for 1 mark (sand dunes or mangroves would be acceptable also) as is the answer to Part (b)(i), which gains 2 marks because the working as well as the correct answer is shown. Mark schemes for a 'calculate' question like this will usually mark as correct a range of ±5% from the precise answer. Part (b)(ii) scores 3 marks. This is a good answer that takes the explanation of coastal defences and extends it by mentioning two specific types and the impact of groynes on longshore drift. Differences in geology and resistance to erosion could also be used as a valid explanation.

(c) Explain the formation of **two** coastal depositional landforms. (4 marks)

(d) Explain why it is difficult to predict future sea levels. (6 marks)

ⓔ Part (c) needs two different landforms, with an extended explanation for each one. Be careful to choose depositional landforms not erosional ones such as arches or stacks. There are no marks available in Part (c) for describing the landform, only for explaining the processes that formed it. Part (d) is a Levels-marked question (see Levels mark scheme on page 90). A range of explanations is needed that might explain why predicting sea levels in the future is difficult. These could be either physical processes or human issues.

Student answer

(c) Spits are formed by the process of longshore drift as sediment is moved along-coast by the action of waves with net transport in one direction. Wave erosion at the end of the spit is balanced by sand deposition stabilising the spit. Where two longshore currents meet from different directions a cuspate foreland may form as the opposing currents cancel each other out leading to deposition on either side of the V-shaped sand bar.

(d) Sea levels are hard to predict for two main reasons. Firstly, it is hard to know how ice stores will respond to global warming. Small differences in future global temperatures could result in large changes to melting rates on Greenland and Antarctica. Most sea level rise between 1990–2010 was due to thermal expansion increasing the volume of ocean water. Future thermal expansion depends on future temperature which can't be known. In some locations such as Norway and Scotland isostatic change is happening alongside eustatic sea level rise making prediction complex. Secondly, because future sea levels depend on future temperature, predictions are only possible if future temperatures are known. They are not because they depend on future greenhouse gas levels and the amount of these depends on other unknowns such as future global population, affluence, fossil fuel use as well as attempts to reduce emissions.

🄔 **10/10 marks awarded.** The answer to Part (c) scores 4 marks. Two different landforms are explained with good use of terminology. The answer avoids describing the landforms and instead focuses on processes such as longshore drift. In each case two explanations are provided. This type of extended explanation is needed to gain 2×2 marks.

The answer to Part (d) scores Level 3, 6 marks. It focuses on the uncertainty of future sea levels in terms of physical processes such as ice melt and human activities that could affect the rate of future planetary warming. The range of explanations is good, as is the use of terminology which is precise and shows understanding (eustatic, isostatic, thermal expansion). There is understanding of the links between human actions and physical systems and some located examples are used to support the explanation.

(e) Assess the degree to which coastal management decisions can result in conflict at the coast. (12 marks)

🄔 This extended question uses the command word 'assess', which means 'weigh-up'. The key here is to demonstrate understanding of a range of situations where policy decisions have led to conflict between different groups, using examples to support your answer (evidence) and then make a judgement about how likely conflict is. Answers must focus on situations where there is some conflict but also examples of where consensus has been achieved. Level 3 marks are achieved by making supported judgements about how likely conflict is. Answers that just explain different management decisions and polices are likely to get no more than 8 marks. Successful arguments would include those that argue conflict is more likely in some places, or because of some policies, compared with others. Questions in this format do not have a 'right' answer; it is the quality of your assessment and judgements that gain marks.

Student answer

(e) Coastal decision making involves a wide range of groups and decision makers and can often lead to conflict. Conflict is almost inevitable when No Active Intervention or Strategic Realignment polices are chosen b. This is because some property is likely to be lost. This is the case at Happisburgh on the North Norfolk coast where 20–35 properties are likely to be lost to erosion by 2105 a and the current policy is No Active Intervention. Homeowners, who will receive no compensation, are in conflict with North Norfolk as well as UK government policy d. At Hornsea on the Holderness Coast the decision to protect the town with sea-walls, groynes and flood walls (Hold the Line policy) was welcomed by residents and business owners but not by farmers downdrift a. Hornsea's groynes trap longshore drift sand, starving beaches downdrift of sand and increasing erosion there c. This example shows that protecting one location affects others because of the interconnected nature of the coastal sediment cell d. On the other hand sometimes a decision can be reached that pleases all stakeholders. Abbotts Hall Farm on the Blackwater Estuary in Essex was bought in 2000 by Essex Wildlife Trust a e. A 4,000 hectare managed realignment scheme was implemented by creating five breaches in the sea-wall in 2002 b. This allowed new salt marshes to form inland. This benefited the Abbots Hall Farm owners who received the market price for their threatened farm. Local government avoided the costs of a Hold the Line policy and environmentalists were pleased with the sustainable approach that protected wading bird habitats by expanding the salt marsh area c. Since 2002 the area has become more accessible for recreation as well as having the risk of flooding reduced. Policy decisions do not inevitably lead to conflict f.

e **12/12 marks awarded.** This is a Level 3 answer. a The answer is supported by a range of examples with some good factual detail. b There is good understanding of a range of policy decisions that are used in the UK. c These are related to physical processes such as longshore drift which shows wider understanding of the coastal system. d Evidence is presented to support the conflict idea, e but crucially an alterative example is provided that argues that conflict does not always occur. This provides a balanced argument. f There is a clear judgement that supports the evidence presented.

■ A-level questions

Tectonic processes and hazards

Question 1

(a) Study Figure 3. For each location shown in Figure 3, calculate the tsunami travel-time from the origin to the location.

(4 marks)

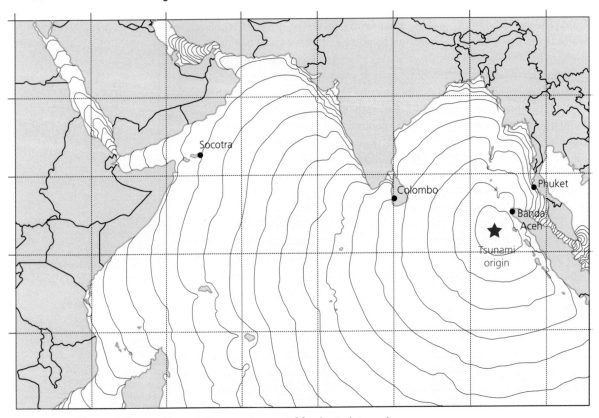

Figure 3 Tsunami travel-time contours at 30 minute intervals

(b) Assess the importance of prediction and forecasting in the successful management of tectonic disasters.

(12 marks)

🅔 Part (a) is a skills question, using the information on Figure 3. The key here is to be accurate and to recognise that the isolines on Figure 3 are in 30 minute, not 1 hour, intervals. Don't rush these skills or 'doing' questions as mistakes are easy to make.

Part (b) is an extended writing question that is marked in Levels. 'Assess the importance' means 'weighing up' so your answer needs to be evaluative in style. Both prediction and forecasting need to be mentioned, and high quality answers will also consider other factors such as response and evacuation, as prediction and forecasting are not the only factors that determine if tectonic disasters are successfully managed. Examples need to be used to support your answer to give it the required depth. The

Levels mark scheme for 12-mark questions is shown below. The same Levels mark scheme is used for the 12-mark questions in the AS options questions.

Level 1 1–4 marks	■ Demonstrates isolated elements of geographical knowledge and understanding, some of which may be inaccurate or irrelevant ■ Applies knowledge and understanding of geographical information/ideas, making limited logical connections/relationships ■ Applies knowledge and understanding of geographical information/ideas to produce an interpretation with limited relevance and/or support ■ Applies knowledge and understanding of geographical information/ideas to make unsupported or generic judgements about the significance of few factors, leading to an argument that is unbalanced or lacks coherence
Level 2 5–8 marks	■ Demonstrates geographical knowledge and understanding, which is mostly relevant and may include some inaccuracies ■ Applies knowledge and understanding of geographical information/ideas logically, making some relevant connections/relationships ■ Applies knowledge and understanding of geographical information/ideas to produce a partial but coherent interpretation that is mostly relevant and supported by evidence ■ Applies knowledge and understanding of geographical information/ideas to make judgements about the significance of some factors, to produce an argument that may be unbalanced or partially coherent
Level 3 9–12 marks	■ Demonstrates accurate and relevant geographical knowledge and understanding throughout ■ Applies knowledge and understanding of geographical information/ideas logically, making relevant connections/relationships ■ Applies knowledge and understanding of geographical information/ideas to produce a full and coherent interpretation that is relevant and supported by evidence ■ Applies knowledge and understanding of geographical information/ideas to make supported judgements about the significance of factors throughout the response, leading to a balanced and coherent argument

Student answer

(a)

Location	Tsunami travel-time	
	Hours	Minutes
Banda Aceh	1	30
Colombo	2	30
Phuket	2	30
Socotra	6	30

(b) Prediction means the ability to say when and where a hazard will strike. For prediction to be useful it has to be accurate to within a few days, otherwise evacuations based on predictions become lengthy and problematic. Forecasting means giving the percentage chance of a hazard occurring a. For earthquakes, forecasting is possible c. For instance the USGS forecasts that many California locations have a 10% chance of a 6.7 earthquake in the next 30 years b. This type of forecast quantifies risk, so is useful for people buying homes or for emergency services planning for a disaster. Earthquakes cannot be predicted, so prediction is of no use. Far more important is preparation and public education so people know what to do during and after an earthquake. Land-use zoning can be used to avoid building in area of high risk, e.g. risk of liquefaction, and this can reduce economic and human losses. In contrast, volcanic eruptions can increasingly be predicted using gas spectrometers, tiltmeters and by analysing the seismic 'noise' made my mobile magma **c**. This allows for timely warning and evacuation. For instance 10,000s of people were saved by the timely evacuation around Mt Pinatubo when it erupted in 1991 **b**. However, monitoring volcanoes is costly and not done in all cases. This means that education, preparation and response should a disaster occur are important **e**. Technology such as ocean monitoring buoys can be used to predict the arrival of tsunami **c**, and even 5–10 minutes warning can save large numbers of lives. However, it cannot prevent very high economic losses such as the US$150 billion in losses caused by the 2011 Sendai tsunami in Japan **b**. In conclusion, prediction is very important for eruptions and tsunami, but only forecasting is possible for earthquakes **d**. In all cases, prediction and forecasting tends to reduce human losses. Reducing economic loss depends more on long preparation and response.

e **16/16 marks awarded.** Part (a) scores 4 marks as all of the answers are accurate. There is some leeway in the minutes, but not the hours. The answer to Part (b) is very good and scores 12 marks. It is well structured and uses good terminology. **a** It shows a good understanding of both forecasting and prediction, recognising them as different things. **b** There are examples used to support the answer, which provides depth. **c** The answer is applied to volcanoes, earthquakes and tsunami so a good range of hazards is considered. **d** The value of prediction is considered and judgements are made contrasting the case of volcanoes versus earthquakes. **e** There is also assessment by considering that even though prediction is useful for eruptions it is not available everywhere, meaning that other factors are important if disasters are to be successfully managed.

Landscape systems, processes and change: Glaciated landscapes and change

Question 2

(a) Study Figure 4. **Explain the evidence for recent glacier retreat in Figure 4.** (6 marks)

e This is a data stimulus question. The resource, in this case a photograph, needs to be studied carefully and evidence from it used to answer the question. The photograph shows a small glacier snout (lower, centre) with an extensive area of light-coloured sediment in front and at the side, with a wide and deep glacial valley in the background. These 6-mark questions in the Landscape systems, processes and change options are marked using the following Levels mark scheme.

Figure 4 The snout of the Höllentalferner glacier in Germany

Level 1 1–2 marks	■ Demonstrates isolated or generic elements of geographical knowledge and understanding, some of which may be inaccurate or irrelevant ■ Applies knowledge and understanding to geographical information inconsistently. Connections/relationships between stimulus material and the question may be irrelevant
Level 2 3–4 marks	■ Demonstrates geographical knowledge and understanding, which is mostly relevant and may include some inaccuracies ■ Applies knowledge and understanding to geographical information to find some relevant connections/relationship between stimulus material and the question
Level 3 5–6 marks	■ Demonstrates accurate and relevant geographical knowledge and understanding throughout ■ Applies knowledge and understanding to geographical information logically to find fully relevant connections/relationships between stimulus material and the question

Questions & Answers

> **Student answer**
>
> **(a)** In Figure 4 a small glacier can be seen in the centre at the bottom. It is surrounded on either side and in front by light-coloured sediment. This is likely to be glacial moraine. It is recently formed because it is not covered in vegetation, unlike the valley floor in the distance. There are lines and ridges on the right in the moraine, which could be recent terminal moraine lines suggesting the glacier was larger in the recent past. This glacial sediment has been dumped as the glacier has retreated towards the bottom of the photo. The steep rock valley sides on the left suggest the glacier was once much higher and filled the valley. Very bare, rough rock on the valley sides suggests recent glacial abrasion and plucking to create the craggy rock sides. The valley in the distance where the vegetation appears is roughly U-shaped, which could be explained by the ice filling the valley and extending over that area in the past. There is also an area that looks party vegetated but has no trees. This is probably the moraine from 50–100 years ago which has had time for some plants to grow in some areas.

@ **6/6 marks awarded.** The strength of this answer is its careful observation of the photograph. Features such as bare sediment, vegetated and partially vegetated areas are identified and related to possible past positions of the glacier. There is some interpretation and explanation of moraines, a key landform in identifying former ice positions, as well as an explanation of the shape of the valley and the possible past position of glacier ice. Good understanding of glacial erosion processes is shown in terms of their impact of the valley shape and features.

(b) Explain how glacial landforms can be used to help reconstruct former ice mass movement and extent.

(8 marks)

@ This style of question is best thought of as a 'mini-essay'. The topic is quite a narrow one, focused on a small area of physical geography. There is no stimulus material so you need to use detailed knowledge and understanding to provide explanations. Examples, but not large case studies, can also be used to support your explanations. In this question answers need to explain both ice mass extent and movement. These 8-mark questions in the Landscape systems, processes and change options are marked using the following Levels mark scheme.

Level 1 1–2 marks	▪ Demonstrates isolated elements of geographical knowledge and understanding, some of which may be inaccurate or irrelevant ▪ Understanding addresses a narrow range of geographical ideas, which lack detail
Level 2 3–5 marks	▪ Demonstrates geographical knowledge and understanding, which is mostly relevant and may include some inaccuracies ▪ Understanding addresses a range of geographical ideas, which are not fully detailed and/or developed
Level 3 6–8 marks	▪ Demonstrates accurate and relevant geographical knowledge and understanding throughout ▪ Understanding addresses a broad range of geographical ideas, which are detailed and fully developed

Student answer

(b) Ice masses include glaciers and ice sheets. The former extent of large ice sheets can be partly indicated by isostatic readjustment **a** of the land surface which was depressed by the weight of ice. This method is only useful for large ice sheets not valley glaciers **c**. Far more accurate is the use of terminal moraines **b**. These mounds of glacial sediment can indicate the maximum extent of glacier and ice sheets. The Valparaiso moraine south of Chicago indicates the maximum southward extent of last glacial period. Other landforms, such as eskers **b**, are known to have formed englacially or subglacially **a** so can indicate area that had ice cover. Eskers may also indicate flow direction assuming meltwater and ice flowed in similar directions. Large ice sheets have a meltwater plain or sandur in front of the ice **b** which consists of meltwater channels and water deposited sediment. The edge of this sandur can be used to infer the edge of the ice. One of the most useful indicators of ice flow direction are erratics **b**. These are rocks and boulders deposited in a location distant from their source rock because the rocks were carried by ice. If the source outcrop can be identified then the flow direction of ice can be inferred. Striations **b**, which are grooves cut into rock by abrasion **a**, also indicated ice flow direction but only over short distances **c**.

e **8/8 marks awarded.** This is a Level 3 answer. **a** It uses good terminology about glacial processes. **b** A range of different landforms are considered and their role is explained. The answer relates to both ice extent and ice movement and has more than one example of a landform for each so the answer has a good range of ideas. **c** There are some comments on the usefulness of the landforms, which, while not directly asked for, help demonstrate depth of understanding.

(c) Evaluate the extent to which management can balance the demands of conservation and economic development in glacial environments. *(20 marks)*

e This question is an essay question with a high mark tariff. The question has a number of different elements to it (conservation + economic development) all of which need to be covered. There are concepts which need to be addressed too (management, demands). The command word 'evaluate' means 'weigh-up and come to a judgement'. Good answers will argue a case using examples and case studies to back up the argument. For this question, it might be argued that the balance is possible to achieve, but that is not the case everywhere, i.e. in areas with high pressure versus more isolated areas with few inhabitants or visitors. These 20-mark questions in the Landscape systems, processes and change options are marked using the following Levels mark scheme.

Level 1 1–5 marks	■ Demonstrates isolated elements of geographical knowledge and understanding, some of which may be inaccurate or irrelevant ■ Applies knowledge and understanding of geographical ideas, making limited and rarely logical connections/relationships ■ Applies knowledge and understanding of geographical information/ideas to produce an interpretation with limited coherence and support from evidence ■ Applies knowledge and understanding of geographical information/ideas to produce an unsupported or generic conclusion, drawn from an argument that is unbalanced or lacks coherence
Level 2 6–10 marks	■ Demonstrates geographical knowledge and understanding, which is occasionally relevant and may include some inaccuracies ■ Applies knowledge and understanding of geographical information/ideas with limited but logical connections/relationships ■ Applies knowledge and understanding of geographical ideas in order to produce a partial interpretation that is supported by some evidence but has limited coherence ■ Applies knowledge and understanding of geographical information/ideas to come to a conclusion, partially supported by an unbalanced argument with limited coherence
Level 3 11–15 marks	■ Demonstrates geographical knowledge and understanding, which is mostly relevant and accurate ■ Applies knowledge and understanding of geographical information/ideas to find some logical and relevant connections/relationships ■ Applies knowledge and understanding of geographical ideas in order to produce a partial but coherent interpretation that is supported by some evidence ■ Applies knowledge and understanding of geographical information/ideas to come to a conclusion, largely supported by an argument that may be unbalanced or partially coherent
Level 4 16–20 marks	■ Demonstrates accurate and relevant geographical knowledge and understanding throughout ■ Applies knowledge and understanding of geographical information/ideas to find fully logical and relevant connections/relationships ■ Applies knowledge and understanding of geographical information/ideas to produce a full and coherent interpretation that is supported by evidence ■ Applies knowledge and understanding of geographical information/ideas to come to a rational, substantiated conclusion, fully supported by a balanced argument that is drawn together coherently

Student answer

(c) There are some examples of glacial environments where conservation of biodiversity and the unique landscape is placed well ahead of desires to develop areas for their economic resources. The most famous example is Antarctica [a] which has been protected under the Antarctic Treaty since 1961 [c]. All commercial exploitation of mineral resources is banned and fishing and tourism are strictly regulated. This works because there is an international consensus that Antarctica [b] is a unique, pristine environment that should be the preserve of science not economic development. Elsewhere views are not so clear cut [e]. The periglacial tundra of Northern Alaska, the Antarctic and ANWR are remote, isolated active glacial environments. Relict [b] glacial environments such as the UK's Lake District [a] are much more accessible and consequently are under pressure from tourism. Tourism contributes up to 40% of the area's economy and developing it creates jobs and economic development. However, fragile uplands are prone to trampling, litter and footpath erosion and the area's carrying capacity may even be exceeded in summer months [e]. The Lake District National Park authority manages the area using the 'Sandford Principle' meaning that conservation takes priority over economic development if the two conflict [c]. While this helps preserve the glacial landscape it risks making people dependent on a small number of low paid, seasonal jobs [d].

It also means large developments such as HEP reservoirs and quarries have no chance of being developed. In the Alps [a], which has active and relict [b] landscapes, the fact that managing the area involves numerous different countries has meant that damage to the landscape from ski resorts, motorways and urbanisation has occurred despite the Alpine Convention that should protect the area. Overall, a balance is easier to strike in extreme, isolated environments with few or no residents. It is much harder in accessible locations and where more than one country is involved [f].

e **20/20 marks awarded.** This answer is Level 4. [a] It uses a wide range of different examples, which are applied to the concepts of economic development and conservation. [b] The range of examples includes both active and relict landscapes, giving good balance. [c] The examples include details of how the areas are managed. [d] There are also judgements made about the success of the different management strategies. [e] The overall approach is evaluative with different views being presented to produce a coherent overview. [f] There is a clear overall judgement that draws together the evidence presented.

Landscape systems, processes and change: Coastal landscapes and change

Question 3

(a) Study Figure 5. Explain the formation of the landforms shown. (6 marks)

Figure 5 Landforms on the Costa Brava, Spain

🅔 This is a data stimulus question. The resource, in this case a photograph, needs to be studied carefully and evidence from it used to answer the question. The photograph shows a small rock arch in a cliff. There is also a cave feature beneath and to the right of the arch. In several places, e.g. around the cave mouth, a small wave-cut notch can be seen. A small rock stump can be seen in the foreground.

> ### Student answer
>
> **(a)** The main landform in Figure 5 is a rock arch. This formed by abrasion and hydraulic action. The rock in Figure 5 looks like heavily fractured limestone. Hydraulic action would happen when incoming waves crashed into the cliff face pressurising air in the fractures and forcing them apart. Combined with the sand-papering action of abrasion as waves hurl sediment at the rock, the erosion has forced through a small headland, forming an arch. The top of the arch is protected from direct wave erosion due to its height. To the right of the arch is a cave. This is the first stage of arch formation. The cave could form where a weakness like a small fault or large fracture creates an area of weak rock that abrasion and hydraulic action can exploit. The high temperatures in Spain may mean some chemical weathering has also contributed to the landforms, e.g. by widening joints and fractures.

🅔 **6/6 marks awarded.** This answer uses good terminology to explain the erosion processes that formed the arch, and makes good use of Figure 5. Observations from the photograph are accurate and the answer picks out and explains details of the rock type and smaller landforms such as the cave. There is detailed understanding of the impact of weaknesses in the rock, and how abrasion and hydraulic action contribute to the formation of both the arch and cave. Some wider understanding is shown by the suggestion that weathering would also a contributing factor.

(b) Explain how bedrock lithology and geological structure can influence rates of coastal recession. (8 marks)

🅔 This style of question is best thought of as a 'mini-essay'. The topic is quite a narrow one focused on a small area of physical geography. There is no stimulus material so you need to use detailed knowledge and understanding to provide explanations. Examples, but not large case studies, can also be used to support your explanations.

> ### Student answer
>
> **(b)** Lithology has a significant impact on rates of coastal recession. Resistant rocks often form headlands. An example is Flamborough Head in Yorkshire [a]. The chalk here erodes at 1–2 mm per year through abrasion and hydraulic action [b]. South of Bridlington the boulder clay of the Holderness Coast erodes at 2–3 m per year because it is much less resistant to erosion [b]. Boulder clay also recedes due to mass movements, especially rotational slides. These are often linked to storm events which undermine the cliffs but also saturate them with heavy rain, leading to internal failure. Lulworth Cove shows how rock type and structure can influence [a]. The cove is part of

the concordant Dorset coast [c]. Resistant beds of Portland and Purbeck limestone form the narrow cove entrance, and hard chalk the steep cliff at the back of the cove [b]. The wide part of the cove has been eroded from softer clays, which are less resistant to erosion. On the Northumberland coast rates of recession are often determined by the weakest strata in a cliff [a]. Often this is a coal seam, which readily forms a wave-cut notch, leading to the collapse of more resistant rocks above [b]. Faults and large joints are often preferentially eroded as they are weaker than surrounding rocks, and become the locations for caves and arches eroding more quickly than surrounding rocks [d].

@ **8/8 marks awarded.** This is a Level 3 answer. [a] The answer uses a number of real examples that add depth to the explanation. [b] The importance of lithology (rock type) is explained with reference to these examples and some data are provided on recession rates. [c] The Lulworth Cove example combines rock type and geological structure, showing that the two cannot be dealt with separately. [d] Some further details of minor structural features are also explained. The answer has a good range of ideas for both lithology and structure.

(c) Evaluate the extent to which the twin threats of sea level rise and erosion on coastlines can be managed in a sustainable way.

(20 marks)

@ This question is an essay question with a high mark tariff. The question has a number of different elements to it (sea-level rise + erosion + management) all of which need to be covered. There are concepts that must be addressed too (threat, sustainable). The command word 'evaluate' means 'weigh-up and come to a judgement'. Good answers will argue a case using examples and case studies to back up the argument. For this question it might be argued that erosion can be managed in a sustainable way as it is a local threat, whereas sea level rise is much harder to manage as sea level is rising everywhere (although it threatens some coasts much more than others).

Student answer

(c) Sustainable management of coasts means managing littoral cells holistically rather than managing isolated pockets of coast [a]. It also means management that meets the social and economic needs of residents and users as well as maximising environmental protection. Erosion threats can be managed using Integrated Coastal Zone Management (ICZM) and Shoreline Management Plans (SMPs) [b]. In the UK SMPs are drawn up by local authorities working collaboratively, and with all stakeholders, to manage a long stretch of coastline. Decisions about which areas to protect from erosion, i.e. Hold the Line, are made based on the cost and technical feasibility of defences versus the value of what is being protected. On the Holderness coast towns like Hornsea and Withernsea are protected but low value farmland is not [c]. The overall SMP works with the natural Holderness sediment cell to maximise protection in some places with the overall aim of reducing erosion rates over a timescale of 100 years or more [d].

Erosion management is fairly local in scale and the threatened areas tend to be small ⓔ. Sea level rise is a widespread threat and as such is much harder to manage. It threatens London, as well as the Essex Marshes, large areas on Norfolk and Lincolnshire and many other areas ⓒ. It cannot be managed at a local level only because the basic cause of most sea level rise is global warming. This problem requires action at a global scale to reduce emissions. Local areas are left to deal with the consequences of sea level rise but cannot manage the fundamental cause. However, the threat can be managed in a sustainable way to some extent. The case of Abbot's Hall Farm on the Blackwater Estuary in Essex shows that numerous stakeholders can agree even when difficult decisions need to be made ⓕ. In 2002 five breaches in the sea-wall here turned 4000 hectares of land into a managed realignment scheme. This scheme removed the risk of sea level rise by turning a once protected areas in to new salt marshes showing that environmentalists, landowners, coastal managers and local people and businesses can all be kept happy even when radical plans are adopted. Overall, managing coastal erosion in a sustainable way is possible if it is done in a holistic way. There will always be winners and losers but decisions can ensure most people are winners. Sea level rise is more about managing consequences than causes and as such is harder, but even radical new ideas can lead to sustainable outcomes ⓖ.

ⓔ **20/20 marks awarded.** This is a Level 4 answer. It shows good conceptual understanding by ⓐ defining sustainable management at the start and ⓑ relating this to specific strategies. ⓒ A range of real examples are used with some detail, which provides the evidence needed to make a judgement. ⓓ The Holderness example is used to show that if management is holistic it can be sustainable. There is a good balance between the sections on erosion and sea level rise. ⓔ There is an evaluative comparison made between the two threats, contrasting the extent to which they can be managed. ⓕ The Blackwater Estuary case study is used very well to show that even the more difficult problem of sea level rise can be managed sustainably. ⓖ The overall conclusion is supported by the evidence presented.

Knowledge check answers

1 Oceanic plates are more dense than continental plates, and oceanic plates are thinner than continental plates.

2 Scientists have no direct observations of tectonic processes occurring inside the Earth, so it remains a theory.

3 An earthquake is a release of stored energy, which travels through the Earth as a series of waves.

4 Convergent boundaries/subduction zones produce explosive eruptions with multiple hazards.

5 The sea bed has to be displaced vertically, either up or down.

6 A natural disaster.

7 They are logarithmic scales (non-linear).

8 Subduction zones (destructive plate margins) can generate earthquakes of magnitudes greater than 9.0.

9 Landslides are very common, especially in mountainous areas.

10 Owing to the higher population density in urban areas, meaning more vulnerable people.

11 Numbers affected.

12 High magnitude, rare events, impacts on more than one country

13 No. The level of risk can be forecast, but precise times and locations (a prediction) are not possible.

14 The recovery stage.

15 Modify the loss, because loss only occurs after the disaster.

16 The orbital eccentricity cycle, lasting 100,000 years, has the largest impact.

17 The Little Ice Age cold phase is usually linked to the Maunder Minimum.

18 The Devensian glacial period.

19 Warm-based glaciers have water at their base.

20 The Antarctic ice sheet is the largest single ice mass on Earth.

21 Freeze–thaw, or frost-shattering, weathering.

22 Firn is the intermediate stage in ice formation, between freshly fallen snow (neve) and ice.

23 Accumulation.

24 The increase in global average air temperature caused by global warming.

25 Basal slip (sometimes called basal sliding).

26 Approximately 12,500 years ago.

27 Chattermarks are produced by crushing, whereas striations are produced by abrasion.

28 A pyramidal peak.

29 The steep (crag) sides point in a different direction; upstream for crag and tail and downstream for roche moutonnée.

30 A rock or boulder moved by ice from its area of origin to a new location.

31 Fluvioglacial; because flowing water aligns sediment particles in the direction of flow.

32 Mountains, ridges, passes and lakes.

33 Lichens, mosses and springtails.

34 A glacial outburst flood, most commonly caused by volcanic eruption under an ice mass.

35 Glacial meltwater is an important source of water supply in many parts of the world.

36 It is a worldwide risk, so everywhere is potentially affected by it.

37 A secondary coast.

38 The layers are technically termed strata.

39 Caves are often formed because of weaknesses in the cliff.

40 Sandstone is a clastic rock, formed from cemented sand particles.

41 Plant roots bind sediment together and leaves protect the sediment surface from erosion.

42 Waves begin to break when water depth is roughly half the wavelength.

43 Summer beach profiles are steeper, as they have a depositional berm at the back.

44 Longshore drift.

45 The sediment cell consists of inputs (sources), processes (transfers) and outputs (sinks).

46 Carbonation, hydrolysis and oxidation.

47 Eustatic, isostatic changes are localised.

48 A ria, or a fjord.

49 The IPCC suggest between 28 and 98 cm by 2100; many scientists consider 1 metre likely.

50 During major storms.

51 Storm surges are caused by low air pressure.

52 Coastal flooding.

53 Groynes, because their purpose is to trap sediment moving along the coast.

54 A Shoreline Management Plan (SMP).

55 No Active Intervention.

56 To determine whether there is an economic case for the defences, i.e. the benefits outweigh the costs.

Index

Note: **bold** page numbers indicate key terms.

Index